Chemistry Research and Applications

nova
science publishers
www.novapublishers.com

Chemistry Research and Applications

The Future of Biorefineries
Waldemar Nyström (Editor)
2023. ISBN: 979-8-88697-524-6 (Hardcover)
2023. ISBN: 979-8-88697-528-4 (eBook)

The Essential Guide to Alkaloids
Deepak Kumar Semwal, PhD (Editor)
2023. ISBN: 979-8-88697-456-0 (Hardcover)
2023. ISBN: 979-8-88697-526-0 (eBook)

Properties and Uses of Antimony
David J. Jenkins (Editor)
2022. ISBN: 979-8-88697-081-4 (Softcover)
2022. ISBN: 979-8-88697-088-3 (eBook)

The Chemistry of Calcium Carbonate
George R. Hood (Editor)
2022. ISBN: 979-8-88697-442-3 (Hardcover)
2022. ISBN: 979-8-88697-448-5 (eBook)

Selected Topics in Germanium
Charles A. Smith (Editor)
2022. ISBN: 979-8-88697-200-9 (Softcover)
2022. ISBN: 979-8-88697-216-0 (eBook)

The Science of Carbamates
Güllü Kaymak (Editor)
2022. ISBN: 978-1-68507-708-2 (Softcover)
2022. ISBN: 978-1-68507-872-0 (eBook)

More information about this series can be found at
https://novapublishers.com/product-category/series/chemistry-research-and-applications/

Victor Martinez-Luaces
Editor

A Closer Look at Chemical Kinetics

nova
science publishers
www.novapublishers.com

DOI: 10.52305/GLSO9660.

Library of Congress Cataloging-in-Publication Data

ISBN: 979-8-88697-484-3

Published by Nova Science Publishers, Inc. † New York

Contents

Preface

This book proposes a closer look at chemical kinetics, which in this specific case, should be understood as a more detailed study of the connections of this discipline with other branches of chemistry and, even more, with other areas of science and technology.

The first part, which includes two chapters, explores theoretical developments where the connections of chemical kinetics with mathematical modeling of scientific and industrial problems are clearly illustrated.

Chapter 1 focuses on the theoretical analysis of an industrial process with an unknown mechanism. For this purpose, considering that the mathematical structure of a real process does not depend on the measuring system, and using the Gukhman condition for the homogeneity of a function, it is proved that the mathematical model that describes the process can be expressed by power functions of the variables. Besides, this chapter describes an algorithm that allows obtaining the parameters involved in the model.

Chapter 2 studies theoretical aspects of concentration curves in first-order chemical kinetics reactions and mechanisms. The main conclusion is that only four behaviors may occur from an a priori group of seven combinations from a set of three mathematical functions: exponential, trigonometric, and polynomial. The four possible cases are: exponential, exponential-oscillatory, exponential-polynomial, and exponential-oscillatory-polynomial, and all of them are illustrated with first-order chemical mechanisms. The impossibility of the other three behaviors (polynomial, oscillatory, and polynomial-oscillatory) is explained in terms of their eigenvalues, as an application of previous theoretical results.

The second part of the volume is devoted to experimental chemical kinetics, where specific reactions are studied in chapters 3 to 6. In all these chapters, kinetic studies allow the authors to propose mechanisms consistent with the experimental results obtained.

Chapter 3 researches the kinetics of free-radical nonbranched-chain addition processes to several unsaturated compounds. The first example considered is the addition to the C=C double bond of alkenes and their derivatives; the second corresponds to the carbonyl group of formaldehyde, and the third is devoted to the addition of free radicals to one of the multiply bonded atoms of the oxygen molecule. Three reaction schemes are proposed, one for each of those previous cases. A fourth scheme is suggested for the addition of the hydrogen atom to the oxygen molecule and finally, a general scheme for the nonbranched-chain addition to molecules of alkenes, formaldehyde, and oxygen is included. This last scheme summarizes the previous cases and explains the peaks observed experimentally, which arise from competition between saturated and unsaturated component reactions. In all cases, low-reactive free radicals, which do not participate in chain propagation, appear.

Chapter 4 aims to analyze the oxidation of metformin using permanganate in alkaline media. The experimental results show 1:6 stoichiometry (metformin: permanganate) and this oxidation reaction is first-order dependence in permanganate ion and less than unit order in MET and hydroxyl ion concentrations, respectively. The authors propose a mechanism that is reliable with kinetic and spectral observations, and they also study the effect of the ionic strength and the dielectric constant of the reaction media. The various constants involved in different steps of the mechanism are reported. The activation constants and thermodynamic values are also evaluated by a graphical method and discussed in detail.

Chapter 5 investigates a new mechanism for both the nonbranched-chain addition of free radicals to the oxygen molecule and the addition of the hydrogen atoms to the multiple bonds of the oxygen, again in the O_2 molecule. In both cases, the unsaturated compound (i.e., the O_2 molecule) is a reactant and an autoinhibitor. This fact can be explained since O_2 is a source of low-reactive free radicals ($RO_4^{\bullet}$ and $HO_4^{\bullet}$), which act to shorten chain prop-agation. The author presents a scheme involving the radical $RO_4^{\bullet}$ and another one involving $HO_4^{\bullet}$ and, unlike the conventional reaction schemes, they explain the peaks observed in the corresponding experimental curves. The chapter also considers other aspects such as the UV absorbing reactions in the Earth's upper atmosphere and the determination of the kinetic constants involved in the mechanism.

Chapter 6 investigates the oxidation of metformin (MET) with diperiodatocuprate (III) (DPC) in aqueous alkaline media. The experimental results show that the reaction is first order in Cu (III) and less than unit order dependence in MET and OH^-. In this work, the authors propose a mechanism consistent with spectral and kinetic observations, as well as the reaction products. The experiments allow the authors to obtain the kinetic constant and several equilibrium constants for different temperatures, in order to determine activation parameters and thermodynamic quantities.

In the third part of the book, which includes the following two chapters, different types of catalytic processes are studied, particularly those related to electrocatalysis, photocatalysis, and surface phenomena.

Chapter 7 provides an eclectic review that goes from electrochemistry to quantum physics, from chemical kinetics to mathematics, and even industrial and environmental problems. The chapter is divided into three sections: (i) fundamentals of electrode kinetics; (ii) properties of electrocatalysis; and (iii) description of one electrocatalytic process with applications: methanol oxidation. Among other topics, Tafel's law is explained in two different ways, and some criteria are proposed for selecting the best possible electrode –both for the laboratory and industry– and finally, a specific reaction is studied. Different methodologies utilized in this research (FTIRS, DEMS, EMIRS, among others), are described in detail.

Chapter 8 studies semiconductors, photocatalysis, *RedOx*, surface processes, and their interactions. In this research, light-induced electron transfer from Bi_2WO_6 to an electron acceptor in the electrolyte solution, and this process is explained using the interaction between the conduction band potential (ECB) and the density of states (DOS) of the *RedOx* couple in the electrolyte. Energy diagrams illustrate the thermodynamic and kinetic concepts involved in ECB-DOS interactions. The arguments discussed in the chapter contribute to understanding the interplay of physical and chemical phenomena that condition *RedOx* reactions on illuminated semiconductor surfaces.

The fourth part of this volume presents a work related to nanotechnology and its catalytic applications.

In Chapter 9, the authors discuss the effect of graphene oxide (GO) and functionalized graphene oxide (FGO) –using formamidine sulfinic acid (FSA) for this purpose– as an antibacterial product. These products were synthesized for their effective catalytic utilization in antibacterial activity studies against gram-positive and gram-negative bacteria, especially for Streptococcus mutans and *Pseudomonas* bacteria. In the study, different parameters are optimized in order to achieve maximum activity. Finally, a possible explanation and a mechanism are proposed for the observed antibacterial activity.

The fifth and last part of this volume exemplifies the link between chemical kinetics and industrial processes, particularly in metallurgy.

In Chapter 10, the main objective is to analyze the treatment of a residue from the solids collection system at steelworks, which contains relevant amounts of iron, zinc, and lead among minor components such as chromium, tin, nickel, and cadmium. This process includes a hydrometallurgical dissolution with the selective extraction of Zn and Pb, leaving iron compounds in an insoluble form. After the dissolution step, two streams are obtained: a liquid and a solid. The liquid stream is a solution rich in zinc, from which Zn will be electrochemically recovered as a valuable product. The solid stream is stabilized by incorporating it into an alkali-activated inorganic polymer (geopolymer) to obtain a product or residue that can be categorized as non-hazardous. The work focuses on the electrochemical deposition step, its kinetics, the characteristics of the product obtained, and how this knowledge can be used in the design of the corresponding industrial process.

These five sections, each one represented by one to four chapters, clearly illustrate the wide range of applications of chemical kinetics and its connections with other scientific and technological disciplines.

Acknowledgments

I would like to thank the staff of Nova Science Publishers for their support, particularly Thomas Mazzaferro who was very helpful throughout the development and production of this volume.

I also wish to thank Marjorie Chaves who assisted in the editing process and for overseeing the English editing of this book.

Chapter 1

A New Approach to Modeling and Simulation of Industrial Processes

Christo Boyadjiev*
Institute of Chemical Engineering,
Bulgarian Academy of Sciences, Sofia, Bulgaria

Abstract

This chapter presents a theoretical analysis of methods for modeling of the industrial processes kinetics. The role of the kinetics of industrial processes in solving the problems of optimal design and control is analyzed. Thermodynamic, hydrodynamic, and Boltzmann approximations for a mathematical description of the kinetics of industrial processes are described. A new approach to modeling of industrial processes with unknown mechanism, based on an axiom and theorem, is presented. The axiom "A mathematical structure of real process does not depend on the measuring system" is used for the formulation of the theorem "A mathematical structure of a quantitative description of real systems can be presented as a power function complex."

Keywords: modeling, simulation, chemical engineering, industrial processes

Introduction

The main problems in the industry are the optimal design of new devices and the optimal control of active processes, i.e., minimization of the investment and operating costs. These problems are solved by modeling methods [1].

---

* Corresponding Author's Email: chr.boyadjiev@gmail.com.

In: A Closer Look at Chemical Kinetics
Editor: Victor Martinez-Luaces
ISBN: 979-8-88697-484-3

The creation of the mathematical model begins with the formulation of the physical model of the complex process, i.e., the definition of the simple (elementary) processes that make it up and the interactions between them. The second step is to define simple processes that have mathematical descriptions (equivalent mathematical operators). The other simple processes are introduced into the mathematical model through quantitative information obtained from experimental data, which brings the mathematical model as close as possible to the real process. The experiment brings mathematics closer to physics (reality).

The optimal design and control in the chemical industry is uniquely related to processes rates, so all mathematical descriptions of processes are linked to algorithms to determine these rates, i.e., processes kinetics.

Industrial Processes Kinetics

The industrial systems consist of separate phases (gas, liquid, solid) in the industrial apparatuses volumes. They are in thermodynamic equilibrium when the temperatures and concentrations of substances in the individual parts or points of the phases are equal and the velocities are equal to zero.

The processes in the industry are a result of the deviation of the systems from their thermodynamic equilibrium [2]. One system is not in a thermodynamic equilibrium when the velocities, concentrations of the components (substances) and the temperatures at the individual points in the phase volumes are different. These differences are the result of reactions, i.e., of processes that create or consume substance and (or) heat. As a result, the industrial processes kinetics is equivalent to the reactions kinetics [3].

The presented analysis shows that processes in the industry are the result of reactions that occur in the phase volume (homogeneous) or on the boundary between two phases (heterogeneous). Homogeneous reactions are generally chemical, while heterogeneous reactions are chemical, catalytic, physical and chemical adsorption, interphase mass transfer in gas-liquid and liquid-liquid systems (on the interphase surface the substance disappears from one phase and occurs in the other phase). The rates of these processes are determined by the reaction kinetics [3], which lies at the basis of modeling and simulation of the industrial processes.

Modeling

The basics of modeling in chemical engineering, as part of human knowledge and science, are related to the combination of intuition and logic that has different forms in individual sciences [4]. In the mathematics the intuition is the axiom (unconditional statements, that cannot be proven), while the logic is the theorem (the logical consequences of the axiom), but logic prevails over intuition. In the natural sciences (physics, chemistry, biology), the "axioms" (principles, postulates, laws) are not always unconditional, but logic prevails over intuition too.

The processes in the industry take place in the industrial apparatuses, where gas, liquid and solid phases move together or alone. They are described by variables, which are extensive or intensive. In the case of merging of two identical systems, the extensive variables are doubled, but the intensive variables are retained.

In the industry, processes take place in moving phases (gas, liquid, solid). Reactions (reaction processes) lead to different concentrations (and temperatures) in the phase volumes and the phase boundaries. As a result, hydrodynamic processes, diffusion mass transfer and heat conduction are joined to the reaction processes. Under these conditions there are various forms of mass transfer (heat transfer) that are convective (as a result of phase movements) and diffusion (as a result of concentration (temperature) gradients in the phases).

Convective mass transfer (heat transfer) can be laminar or turbulent (as a result of large-scale turbulent pulsations). Diffusion mass transfer (heat transfer) can be molecular or turbulent (as a result of small-scale turbulent pulsations).

Mathematical models of industrial apparatuses aim at determining the concentration of substances (flow temperatures) in the phases. They have different degrees of approximation – thermodynamic, hydrodynamic and Boltzmann's approximations.

Thermodynamic Approximation

The processes in chemical engineering are result of a deviation from the thermodynamic equilibrium between two-phase volumes or the volume and phase boundaries of one phase and represent the pursuit of systems to achieve thermodynamic equilibrium [2]. They are irreversible processes and their kinetics use mathematical structures derived from Onsager's principle of linearity (axiom). According to him, the average values of the derivatives at the time of the extensive variables depend linearly on the mean deviations of the conjugated intensive variables from their equilibrium states. The principle is valid close to equilibrium, and the Onsager's linearity coefficients are kinetic constants. When the process is done away from equilibrium (high intensity processes) kinetic constants become kinetic complexes, depending on the corresponding intensive variables. The thermodynamic approximation models cover the entire volume of the phase or part of it.

Hydrodynamic Approximations

The hydrodynamic level uses the approximations of the mechanics of continua, where the mathematical point is equivalent to an elementary physical volume, which is sufficiently small with respect to the apparatus volume, but at the same time sufficiently large with respect to the intermolecular volumes in the medium. In this level the molecules are not visible, as it is done in the next level of detail of Boltzmann.

The models of the hydrodynamic approximations are possible to be created on the basis of the mass (heat) transfer theory, whose models are created by the models of the hydrodynamics, diffusion, thermal conductivity and reaction kinetics, using the logical structures of three main "axioms," related with the impulse, mass and heat transfer:

1. The postulate of Stokes for the linear relationship between the stress and deformation rate, which is the basis of the Newtonian fluid dynamics models;
2. The first law of Fick for the linear relationship between the mass flow and the concentration gradient, which is the basis of the linear theory of the mass transfer;
3. The first law of Fourier for the linear relationship between the heat flux and the temperature gradient, which is the basis of the linear theories of the heat transfer.

These are the laws of the impulse, mass and energy transfer.

To the above axioms must be added the law of chemical kinetics: the rate of two-molecular reactions depends linearly on the product of the concentrations of the reagents. This law is not an axiom because it follows from the rule of probabilities. The reaction between two molecules takes place when they are at one point. The probability that a molecule of a reagent is at a given point is equal to the ratio of the number of its molecules and the total number of molecules, i.e., it is proportional to its concentration. The same is true for the second reagent. The simultaneous realization of both probabilities is the complex probability, which is equal to the product of the simple probabilities.

Boltzmann's Approximation

In Boltzmann's kinetic theory of the ideal gas, the hydrodynamic "axioms" are three "theorems" that derive from the axiom of the "elastic shock" (in a shock between two molecules the direction and the velocity of the movement change, but the sum of their kinetic energies is retained, i.e., there is no loss of kinetic energy) and the rate coefficients are theoretically determined by the average velocity and the average free run of the molecules.

Mechanism of Influence of Reaction Kinetics

The mathematical model of an engineering process is a mass (heat) balance in the phases volumes, where the mathematical operators are mathematical descriptions of the composite (elementary) processes, and the relationship between them (differential equations) corresponds to the mechanism of the complex process. The boundary conditions of the differential equations are formulated at the interphase boundaries.

Industrial processes are a set of physical and chemical reactions, hydrodynamic, diffusion and thermal processes that take place in the industrial apparatus volume. The problems in the modeling of the kinetics of industrial processes arise from the need for information about the interaction between the simple (elementary) processes in the complex process, i.e., information about the process mechanism.

Modeling of Processes with Unknown Mechanism

There are complex processes whose mechanism is unknown. A typical example of this are the complex chemical reactions, where the rate of which depends on the concentrations of several substances, but the simple chemical reactions among these substances and the relationships among them are unknown.

The kinetics of processes with an unknown mechanism can be modeled on the basis of an axiom, which can be formulated as "The mathematical structure of the quantitative description of real (industrial) processes does not depend on the measuring system of the quantities involved in them".

A. A. Gukhman formulates [5] the condition for homogeneity of a function "The function in the equation $F(x_1, ..., x_n) = 0$ is homogeneous, if is invariant $F(k_1 x_1, ..., k_n x_n) = 0$ with respect to similar transformations $\bar{x}_i = k_i x_i$, $i = 1, ... , n$ ($k_i, i = 1, ... , n$ are positive constants)", i.e.,

$$F(k_1 x_1,...,k_n x_n) = \varphi(k_1,...,k_n) F(x_1,...,x_n). \quad (1)$$

The homogeneous functions satisfy the equation (1) in the cases $F(x_1, ..., x_n) \neq 0$, too.

Using the equation (1), Gukhman formulates and proves the theorem [5] "If mathematical structure is invariant with respect to similar transformations, it is possible to be presented as power functions complex". This Gukhman's theorem can be considered as a consequence of the axiom "Mathematical structure of real process does not depend on the measuring system", because the mathematical structure, which is invariant with respect to similar transformations is the mathematical structure of real processes, which does not depend on the measuring system. As a result, it is possible to formulate the theorem "Mathematical structure of the quantitative description of real systems is possible to be presented as power functions complex". This theorem will be proved for cases of complex chemical reactions.

The kinetics of the complex chemical reactions depends on a set of variables. If the velocity of these processes is denoted by y and the values of these variables are $x_1, ... , x_n$, the equation of the kinetic model will have the form:

$$y = f(x_1,...,x_n). \quad (2)$$

This function is a mathematical structure that is retained when changed the measurement system of the variable, i.e., this mathematical structure is invariant with respect to similar transformations [5]:

$$\bar{x}_i = k_i x_i, \quad i = 1,...,n, \tag{3}$$

i.e., f is a homogeneous function:

$$ky = f\left(k_1 x_1,...,k_n x_n\right) = \phi\left(k_1,...,k_n\right).f\left(x_1,...,x_n\right), \quad k = \phi\left(k_1,...,k_n\right). \tag{4}$$

A short recording of (4) is:

$$f\left[\bar{x}_i\right] = \phi\left[k_i\right] f\left[x_i\right]. \tag{5}$$

The problem consists in finding a function f that satisfies Equation (5). A differentiation of Equation (5) with respect to k_1 leads to:

$$\frac{\partial f\left[\bar{x}_i\right]}{\partial k_1} = \frac{\partial \phi}{\partial k_1} f\left(x_i\right). \tag{6}$$

On the other hand,

$$\frac{\partial f\left[\bar{x}_i\right]}{\partial k_1} = \frac{\partial f\left[\bar{x}_i\right]}{\partial \bar{x}_1}\frac{\partial \bar{x}_1}{\partial k_1} = \frac{\partial f\left[\bar{x}_i\right]}{\partial \bar{x}_1} x_1. \tag{7}$$

From (6, 7) follows:

$$\frac{\partial f\left[\bar{x}_i\right]}{\partial \bar{x}_1} x_1 = \alpha_1 f\left[x_i\right], \tag{8}$$

where

$$\alpha_1 = \left(\frac{\partial \phi}{\partial k_1}\right)_{k_i = 1}. \tag{9}$$

The equation (8) is valid for different values of k_i including $k_i = 1$ $(i = 1, \ldots, n)$. As a result, $\overline{x_i} = x_i \,, i = 1, \ldots, n$ and from (8) follows:

$$\frac{1}{f}\frac{\partial f}{\partial x_1} = \frac{\alpha_1}{x_1}, \tag{10}$$

i.e.,

$$f = c_1 x_1^{\alpha_1}. \tag{11}$$

When the above operations are repeated for $x_2, \ldots, x_n$, the homogenous function f assumes the form:

$$f = k x_1^{\alpha_1} \ldots x_n^{\alpha_n} \tag{12}$$

i.e., the function f is homogenous if it represents a power function complex and as a result it is invariant with respect to similarity (metric) transformations. The parameters $k, \alpha_1, \ldots, \alpha_n$ are determined by experimental data of the industrial process velocity.

Parameters Identification

The parameters $k, \alpha_1, \ldots, \alpha_n$ identification in (12) uses experimental data of the industrial process velocity $f_j^{exp}, j = 1, \ldots, N$ for different values of the variables $x_{1j}, \ldots, x_{nj}$, $j = 1, \ldots, N$, where the calculated values of the industrial process velocity are:

$$f_j = k x_{1j}^{\alpha_1}, \ldots, x_{nj}^{\alpha_n}, \quad j = 1, \ldots, N \tag{13}$$

The parameters identification of the model requires its preliminary logarithmization.

$$F_j = \alpha_0 + \alpha_1 X_{1j} + \cdots + \alpha_n X_{nj}, \; j = 1, \ldots, N, \; F = \log f, \\ \alpha_0 = \log k, \quad X = \log x \tag{14}$$

The differences between the calculated and experimental values of the industrial process velocity will be presented as

$$\alpha_0 + \alpha_1 X_{1j},...,\alpha_n X_{nj} - F_j^{\exp} = \varepsilon_j, \quad j = 1,...,N, \quad F_j^{\exp} = \log f_j^{\exp} \tag{15}$$

The solution of the parameters identification problem can be obtained as a minimization of the least-squares function:

$$Q(\alpha_0, \alpha_1, ..., \alpha_n) = \sum_{j-1}^{N} \left(\alpha_0 + \alpha_1 X_{1j}, ..., \alpha_n X_{nj} - F_j^{\exp} \right)^2 \tag{16}$$

Using the condition:

$$\frac{\partial Q}{\partial \alpha_i} = 0, \quad i = 0,1,...,n \tag{17}$$

i.e.,

$$\begin{aligned}
&\sum_{j=1}^{N} 2\left(\alpha_0 + \alpha_1 X_{1j},...,\alpha_n X_{nj} - F_j^{\exp}\right) = 0, \\
&\sum_{j-1}^{N} 2\left(\alpha_0 + \alpha_1 X_{1j},...,\alpha_n X_{nj} - F_j^{\exp}\right) X_{1j} = 0, \\
&\sum_{j-1}^{N} 2\left(\alpha_0 + \alpha_1 X_{1j},...,\alpha_n X_{nj} - F_j^{\exp}\right) X_{2j} = 0, \\
&\ldots \\
&\sum_{j=1}^{N} 2\left(\alpha_0 + \alpha_1 X_{1j},...,\alpha_n X_{nj} - F_j^{\exp}\right) X_{nj} = 0.
\end{aligned} \tag{18}$$

The solution of the equation set (18) permits to obtain the parameters values in the model (12). Determining parameter $\kappa = 10^{\alpha_0}$ by antilogarithmization of parameter α_0 is not recommended due to increasing computational error of α_0.

The equation (13) can be used for this purpose:

$$k = \frac{1}{N}\sum_{j=1}^{N}\frac{f_j^{\exp}}{x_{1j}^{\alpha_1},\ldots,x_{nj}^{\alpha_n}} \qquad (19)$$

The presented algorithm (13-19) permits to use power functions complex (12) as a universal model of the real (industrial) processes.

Conclusion

In the paper is presented a theoretical analysis of the methods for industrial processes modeling.

The role of the kinetics of industrial processes for solving the problems of optimal design and control is analyzed. The thermodynamic, hydrodynamic and Boltzmann approximations for the mathematical description of the kinetics of industrial processes are described.

A new approach to modeling of industrial processes with unknown mechanism is presented on the base of formulated axiom and theorem [6]. The axiom "Mathematical structure of real process does not depend on the measuring system" is used for the formulation of the theorem "Mathematical structure of the quantitative description of real systems is possible to be presented as power functions complex".

An algorithm, for the parameter's identification in the power functions complex model, permits modeling and simulation of industrial processes.

The presented results offer a universal opportunity to determine the rate of industrial processes, which solves three global problems: 1. Optimal design of devices in new productions; 2. Optimal control of the processes in the operation of productions; 3. Sustainability of industrial processes, i.e., conditions under which the rate of the processes is constant and is not affected by accidental disturbances.

References

[1] Boyadjiev, C. *Theoretical Chemical Engineering. Modeling and Simulation*; Springer-Verlag: Berlin Heidelberg, 2010.

[2] Keizer, J. *Statistical Thermodynamics of Nonequilibrium Processe.* Springer-Verlag: New York, NY, 1987.

[3] Boyadjiev, C. Reaction Kinetics in Chemical Engineering, *Bulgarian Chemical Communications*, 2018, vol. 50(Special Issue-K), 112-119.

[4] Boyadjiev, C., Some Thoughts on Logic and Intuition in Science and Chemical Engineering, *Open Access Library Journal*, 2014, vol. 1, no. 6, 1-5.

[5] Gukhman, A. A. *Introduction to the Theory of Similarity*, Ed. "Higher School": Moscow, 1973.

[6] Boyadjiev Chr., A New Approach to Modeling and Simulation of Industrial Processes, *Journal of Engineering Thermophysics*, 2021, Vol. 30, No. 2, 1–6.

Chapter 2

Concentration Curves in First-Order Chemical Kinetics: All Possible Cases

Victor Martinez-Luaces[1,2*]

[1] Chemical Engineering Institute,
Faculty of Engineering, UdelaR, Montevideo, Uruguay
[2] ProfeSTEAM Project,
University of Granada, Granada, Spain

Abstract

There exists a strong relationship between chemical kinetics and mathematics. For instance, as Zambelli remarks, the first differential equation in chemistry probably appeared in a chemical kinetics paper.

In this chapter, the concentration curves corresponding to the first-order chemical kinetics reactions and mechanisms are studied. In those systems, a plethora of solutions can take place, whereas others are not possible.

In order to study the form and qualitative behavior of the solutions, several previous results –published in chemistry books, journal papers, and mathematics book chapters– are revisited and some new ones are introduced.

Considering the theoretical results, only four possible concentration curves may occur from an a priori group of seven combinations from a set of three mathematical functions: exponential, trigonometric (sine and cosine) and polynomial. The four possible cases are: exponential, exponential-oscillatory, exponential-polynomial and exponential-oscillatory-polynomial.

All these possible cases are analyzed and illustrated with first-order chemical mechanisms, while the impossibility of the other three

* Corresponding Author's Email: victorml@fing.edu.uy.

In: A Closer Look at Chemical Kinetics
Editor: Victor Martinez-Luaces
ISBN: 979-8-88697-484-3

behaviors (polynomial, oscillatory and polynomial-oscillatory) is explained in terms of their eigenvalues, as an application of previous theoretical results.

Keywords: concentration curves, first order reactions, qualitative behavior

Introduction

There exists a long tradition of connections between chemical kinetics and mathematics [1]. For instance, it is well-known that a first-order chemical reaction (FOCR) can be modeled [2, 3] through a linear ordinary differential equation (ODE). Moreover, as Zambelli observed [4], the first ODE in chemistry probably appeared in a chemical kinetics paper written by Wilhelmy more than 170 years ago [5]. The reaction studied by this author was of the form $A \xrightarrow{k} products$ and the corresponding ODE is $\frac{d[A]}{dt} = -k[A]$, where $[A]$ is the reactant concentration, t is time and k is the kinetic constant of this reaction. Obviously, this first example is a FOCR since the reaction proceeds at a rate that depends linearly upon only one reactant concentration. Other important reactions follow the same law as in the next examples.

The decomposition of hydrogen peroxide into water and oxygen [6]:

$$H_2O_2 \to H_2O + \frac{1}{2}O_2 \tag{1}$$

The decomposition of sulfuryl chloride to give sulfur dioxide and chlorine [7]:

$$SO_2Cl_2 \to SO_2 + Cl_2 \tag{2}$$

The reaction of trimethyl bromomethane with sodium hydroxide [8] to give trimethyl methanol and sodium bromide:

$$(CH_3)_3CBr + NaOH \to (CH_3)_3COH + NaBr \tag{3}$$

Decomposition of dinitrogen pentoxide [9] to give nitrogen dioxide and oxygen:

$$2\,N_2O_5 \to 4\,NO_2 + O_2 \tag{4}$$

Decomposition of ammonium nitrite in aqueous solution [10]:

$$NH_4NO_2 \rightarrow N_2 + 2\,H_2O \tag{5}$$

In addition, other important examples arise when a chemical reaction – which is not of first order– has the same behavior as a first-order one due to an excess of one or more reactants. Two examples of this case are below:

Hydrolysis of methyl acetate –like other hydrolysis of esters– in the presence of mineral acids [11]:

$$CH_3COOCH_3 + H_2O \rightarrow CH_3COOH + CH_3OH \tag{6}$$

Hydrolysis of sucrose to give glucose and fructose:

$$C_{12}H_{22}O_{11} + H_2O \rightarrow C_6H_{12}O_6 + C_6H_{12}O_6 \tag{7}$$

It is important to mention that the rate of this reaction can be written as:

$$r = k[C_{12}H_{22}O_{11}][H^+][H_2O] \tag{8}$$

Then, this is a third-order chemical reaction [12]; however, the concentrations of both the solvent H_2O and the catalyst H^+ are normally constant, and so it can be considered a pseudo-first-order reaction.

A balanced equation for a given chemical reaction indicates the reactants and products, but it does not reveal how the reaction takes place. The reaction mechanism is the process by which a reaction occurs, and if all the steps are first-order reactions, we will call it a first-order chemical kinetics mechanism (FOCKM).

When a FOCKM is modeled mathematically, the result is always a linear ODE system [13, 14, 15]. The same situation occurs with any other system of chemical reactions, if all of them are first-order or pseudo-first-order ones. In next section, we consider several examples of FOCKM (or reaction systems) in order to illustrate different mathematical models and the corresponding solutions that may appear.

Basic Mathematical Models for FOCKMs

In this section, we consider three different examples: opposed, consecutive, and parallel or competitive reactions, and we analyze the concentration curves for each case.

Opposed Reactions

A typical example of this kind of reactions is the Mutarotation of Glucose [16, 17]: $\alpha - Glucose \leftrightarrow \beta - Glucose$, which can be represented as $A \leftrightarrow B$ and the corresponding ODE system (in matrix form) is:

$$\begin{cases} \frac{d[A]}{dt} = -k_1[A] + k_{-1}[B] \\ \frac{d[B]}{dt} = k_1[A] - k_{-1}[B] \end{cases} \tag{9}$$

or,

$$\begin{pmatrix} [A]' \\ [B]' \end{pmatrix} = \begin{pmatrix} -k_1 & k_{-1} \\ k_1 & -k_{-1} \end{pmatrix} \begin{pmatrix} [A] \\ [B] \end{pmatrix} \tag{10}$$

The associated matrix $\mathbb{A} = \begin{pmatrix} -k_1 & k_{-1} \\ k_1 & -k_{-1} \end{pmatrix}$ was called FOCKM matrix in earlier works [13, 14, 15]. In a previous NOVA book chapter [14] it was observed that the matrix eigenvalues are in this case: $\lambda_1 = 0$ and $\lambda_2 = -(k_1 + k_{-1})$.

Moreover, in [14] it was observed that the solutions of the ODE system $[A](t)$ and $[B](t)$ are of the form $c_1 e^{\lambda_1 t} + c_2 e^{\lambda_2 t}$, i.e., a linear combination of exponentials corresponding to the FOCKM matrix eigenvalues.

Consecutive Reactions

Consecutive reactions can be defined as those chemical processes where the product of one reaction can be considered as the initial material for other reactions of the mechanism. Polymerization and chlorination of methane are a couple of interesting examples of this kind of reactions. Particularly, in the last case, several compounds such as CH_3Cl, CH_2Cl_2, $CHCl_3$ and CCl_4 are consecutively formed [18].

The simplest example of a consecutive reaction can be schematized as $A \overset{k_1}{\rightarrow} B \overset{k_2}{\rightarrow} C$, where two irreversible first-order reactions $A \overset{k_1}{\rightarrow} B$ and $B \overset{k_2}{\rightarrow} C$ occur for certain substances A, B and C.

For this FOCKM, the corresponding ODE system is:

$$\begin{cases} \frac{d[A]}{dt} = -k_1[A] \\ \frac{d[B]}{dt} = k_1[A] - k_2[B] \\ \frac{d[C]}{dt} = k_2[B] \end{cases} \tag{11}$$

$$\text{or} \begin{pmatrix} [A]' \\ [B]' \\ [C]' \end{pmatrix} = \begin{pmatrix} -k_1 & 0 & 0 \\ k_1 & -k_2 & 0 \\ 0 & k_2 & 0 \end{pmatrix} \begin{pmatrix} [A] \\ [B] \\ [C] \end{pmatrix} \tag{12}$$

The concentration curves (i.e., the solutions of the ODE system) can be written as:

$$[A] = [A]_0 e^{-k_1 t} \tag{13}$$

$$[B] = \frac{k_1[A]_0}{k_2-k_1}(e^{-k_1 t} - e^{-k_2 t}) \tag{14}$$

$$[C] = [A]_0 + \frac{[A]_0}{k_1-k_2}(k_2 e^{-k_1 t} - k_1 e^{-k_2 t}) \tag{15}$$

It should be noted that these solutions are once again, linear combinations of the exponentials corresponding to the FOCKM matrix eigenvalues:

$\lambda_1 = -k_1, \lambda_2 = -k_2$ and $\lambda_3 = 0$.

Obviously, the previous solutions are nonsensical if $k_1 = k_2$, so this particular case deserves to be deeply analyzed. The FOCKM matrix eigenvectors are $\overrightarrow{v_1} = (0,-1,1)$ (associated with the eigenvalue $\lambda_1 = -k_1$, double) and $\overrightarrow{v_2} = (0,0,1)$, (associated with $\lambda_2 = 0$, simple), so the algebraic multiplicity ($AlgMult$) and the geometric multiplicity ($GeoMult$) are not the same for the double eigenvalue $\lambda_1 = -k_1$.

Consequently, the ODE system solutions are linear combinations of $e^{0t} = 1$, $e^{-k_1 t}$ and $te^{-k_1 t}$. Then, at least in a first approach, polynomials and exponentials combined (in a linear or non-linear form) may take place.

Competitive Reactions

Parallel or competitive reactions can be exemplified by ethylene oxide production. This chemical compound –a major component in explosives and an important reagent in many chemical processes– may be formed through the partial oxidation of ethylene [19]:

$$2\ C_2H_4 + O_2 \overset{k_1}{\rightarrow} 2\ C_2H_4O \tag{16}$$

At the same time, ethylene may also be involved in the following combustion reaction:

$$C_2H_4 + 3\ O_2 \overset{k_2}{\rightarrow} 2\ CO_2 +\ 2\ H_2O \tag{17}$$

If the first reaction must be selected, then, a silver catalyst is utilized in the oxidation of ethylene, and so, the first kinetic constant k_1 is significantly increased when it is compared to the second one k_2.

The general situation can be schematized as follows: $A \overset{k_1}{\rightarrow} B$, $A \overset{k_2}{\rightarrow} C$, where chemical species A undergoes one of two irreversible first-order reactions to form either species B or species C.

For this new FOCKM, the corresponding ODE system is:

$$\begin{cases} \frac{d[A]}{dt} = -(k_1 + k_2)\,[A] \\ \frac{d[B]}{dt} = k_1[A] \\ \frac{d[C]}{dt} = k_2[A] \end{cases} \tag{18}$$

or:

$$\begin{pmatrix} [A]' \\ [B]' \\ [C]' \end{pmatrix} = \begin{pmatrix} -(k_1+k_2) & 0 & 0 \\ k_1 & 0 & 0 \\ k_2 & 0 & 0 \end{pmatrix} \begin{pmatrix} [A] \\ [B] \\ [C] \end{pmatrix} \tag{19}$$

The FOCKM matrix eigenvalues are the diagonal entries, that is $\lambda_1 = -(k_1 + k_2)$ and $\lambda_2 = 0$, being this last one a double eigenvalue. The corresponding eigenvectors are $\overrightarrow{v_1} = \left(-\frac{k_1+k_2}{k_2}, \frac{k_1}{k_2}, 1\right)$, associated with the simple eigenvalue $\lambda_1 = -(k_1+k_2)$, and $\overrightarrow{v_1} = (0,1,0)$, $\overrightarrow{v_2} = (0,0,1)$, associated with the double eigenvalue $\lambda_2 = 0$. These facts are reflected in the solutions:

$$[A] = [A]_0 e^{-(k_1+k_2)t} \tag{20}$$

$$[B] = -\frac{k_1[A]_0}{k_1+k_2} e^{-(k_1+k_2)t} + \frac{k_1[A]_0}{k_1+k_2} \tag{21}$$

$$[C] = -\frac{k_2[A]_0}{k_1+k_2} e^{-(k_1+k_2)t} + \frac{k_2[A]_0}{k_1+k_2} \tag{22}$$

It should be noted that since the algebraic and geometric multiplicities are the same for the double eigenvalue (i.e., $AlgMult_{\lambda=0} = GeoMult_{\lambda=0}$), then, the solutions are linear combinations of $e^{0t} = 1$ and $e^{-(k_1+k_2)t}$, without including $t.e^{0t} = t$.

Other Possible Cases

It is well known that every system of n first-order linear ODE is equivalent to an n-th order linear equation [20], and their general solution forms [21] can be represented as in Table 1.

In Table 1 it is possible to observe different functions like exponential, polynomial and trigonometric (sine and cosine) and their products and linear combinations.

Table 1. Linear ODE general solution form

Root	Multiplicity	General solution form
r , real	$m = 1$	Ce^{rt}
	$m > 1$	$(C_1 + C_2t + \cdots + C_mt^{m-1})e^{rt}$
$r = a \pm ib$, complex conjugate	$m = 1$	$(C_1 \cos bt + C_2 \sin bt)e^{at}$
	$m > 1$	$(C_1 + C_2t + \cdots + C_mt^{m-1})e^{at} \cos bt +$ $(C_1 + C_2t + \cdots + C_mt^{m-1})e^{at} \sin bt$

Exponentials and polynomials appeared in previous examples, so the next reasonable question is about the oscillatory behavior and whether it is possible or not in FOCKM problems [22].

Oscillating reactions can be exemplified by the iodine clock reaction [23] (oxidation of potassium iodide by hydrogen peroxide) or the famous Belousov-Zhabotinsky chemical reaction [24]. Nevertheless, these equations are not modeled by a linear ODE system.

Therefore, it is important to analyze if oscillating reactions can occur in FOCKM problems. Moreover, it is possible to list seven combinations, multiplying the basic functions: exponential, trigonometric (sine and cosine) and polynomial. The list includes exponential, oscillatory (trigonometric sine and cosine), polynomial, exponential-oscillatory, exponential-polynomial, polynomial-oscillatory and exponential-oscillatory-polynomial.

In FOCKM, some of these solutions can take place, whereas others are not possible. In order to analyze which of them may occur, several previous results are revisited in next section.

Revisiting Several Previous Results

In this section, several results published in books released by NOVA Science Publishers are revisited [13, 14, 15]. The first one is about the eigenvalue $\lambda = 0$, which is always a root of the characteristic polynomial of the FOCKM matrix.

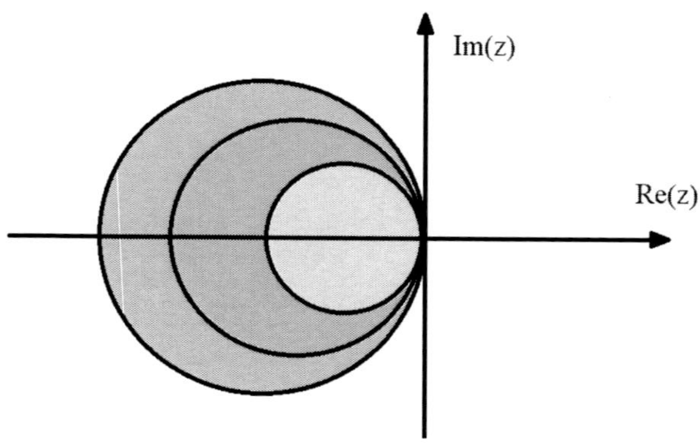

Figure 1. Scheme of a FOCKM Gershgorin Circles.

A more interesting result is about the Gershgorin Circles (i.e., the circles where the matrix eigenvalues can be found) for the FOCKM matrices. Those circles are always tangent to the imaginary axis in the complex plane and the point of tangency is located at the origin. A scheme of these circles can be observed in Figure 1.

Because of the Gershgorin Circles Theorem [25], the system eigenvalues lie always in the union of those circles, and so, all the FOCKM matrix eigenvalues can be found in a circle with radius R, centered in $(-R, 0)$ on the negative part of the real axis.

For instance, in Figure 2 it is possible to schematize the eigenvalues corresponding to several of the previous examples like opposed reactions, competitive reactions, and consecutive reactions in the particular case where $k_1 = k_2$.

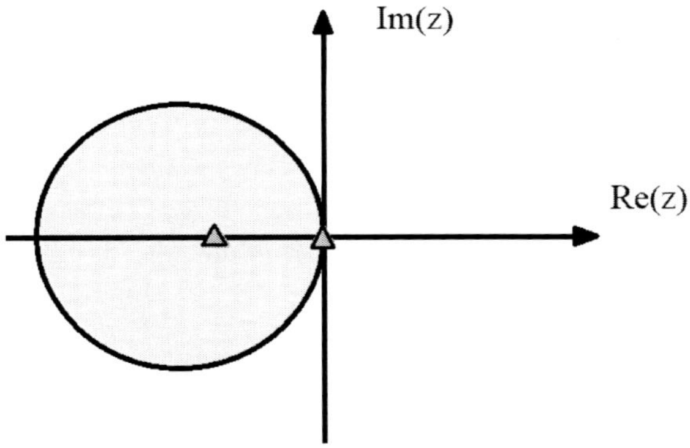

Figure 2. Scheme of the eigenvalues in the previous examples.

A different situation is illustrated in Figure 3, which shows a scheme of the eigenvalues for consecutive reactions with $k_1 \neq k_2$.

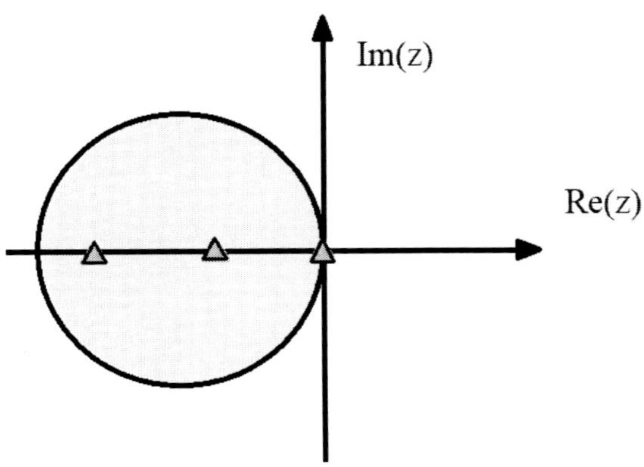

Figure 3. Eigenvalues for consecutive reactions with $k_1 \neq k_2$.

The previous result has several interesting corollaries, including the following:

- If $\lambda = a \pm ib$ are eigenvalues of the FOCKM matrix, then $a \leq 0$.
- $Re(\lambda) = 0$ if and only if $\lambda = 0$.

Then, the real part of a FOCKM eigenvalue is less or equal to zero and only is zero in the case of the null eigenvalue.

In the FOCKM matrix analyzed in the sub-section devoted to competitive reactions it was observed that the eigenvalues are $\lambda_1 = -(k_1 + k_2)$ and $\lambda_2 = 0$, being the last one a double eigenvalue. In the same example, it was noted that the algebraic and geometric multiplicities were the same for $\lambda_2 = 0$, i.e., $AlgMult_{\lambda=0} = GeoMult_{\lambda=0}$. It should be noticed that this is not a particular result, in fact, in [15] it was proved that for every FOCKM the equality $AlgMult_{\lambda=0} = GeoMult_{\lambda=0}$ always happens.

Finally, it is important to remember that for any matrix (FOCKM matrix or not) the inequality $0 = GeoMult_{\lambda} \leq AlgMult_{\lambda}$ holds [26].

Taking into account the previous results, all the possible solutions are listed in the following section.

Possible and Impossible Solutions

As it was observed, the a priori possible solutions form a group of seven different behaviors including: exponential, oscillatory (trigonometric sine and cosine), polynomial, exponential-oscillatory, exponential-polynomial, polynomial-oscillatory and exponential-oscillatory-polynomial.

Exponentials appear when the FOCKM eigenvalues are real, whereas the oscillatory behavior is typical of complex eigenvalues. In order to observe polynomials in the solutions is necessary that $AlgMult_{\lambda} > GeoMult_{\lambda}$, independently of being $\lambda \in \mathbb{R}$ or not.

The analysis of the possible cases is the following:

Case 1: Linear Combination of Exponentials

This is the most typical case in FOCKM, and it happens when $\lambda \in \mathbb{R}$ and $AlgMult_{\lambda} = GeoMult_{\lambda}$. The general solution is a linear combination of the corresponding exponentials, like the product concentration in Equation 15.

Note that among these exponentials, the particular case for $\lambda = 0$ should be considered and since $e^{0t} = 1$, this term is represented by the first addend $[A]_0$ in (15).

Case 2: Pure Oscillatory

Theoretically, this behavior is obtained if $\lambda = 0 \pm ib$ are eigenvalues of the FOCKM matrix, being $b \neq 0$. Nevertheless, this situation is impossible since $Re(\lambda) = 0$ if and only if $\lambda = 0$.

This fact can be observed graphically in Figure 4, where a couple of theoretical eigenvalues $\lambda = 0 \pm ib$ with $b \neq 0$ are represented.

It should be noted that in this hypothetical case, both eigenvalues lie outside the union of Gershgorin Circles. Moreover, this situation is not possible since it contradicts a previous corollary which states that $Re(\lambda) = 0$ if and only if $\lambda = 0$.

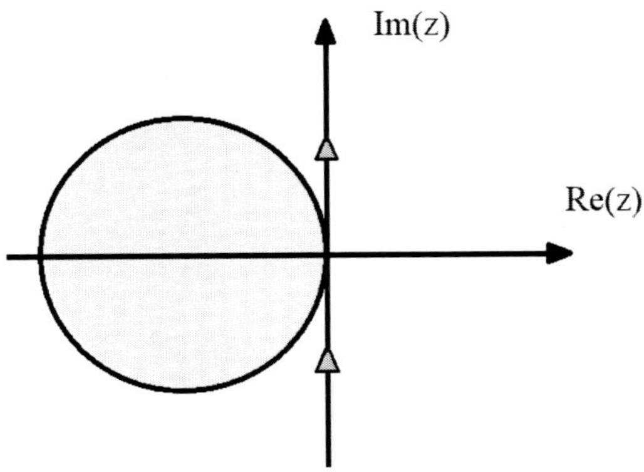

Figure 4. Scheme for hypothetical eigenvalues $\lambda = 0 \pm ib$ with $b \neq 0$.

Case 3: Pure Polynomial

This is another case that it is not possible for FOCKM matrices. In order to be present this kind of solutions, $\lambda = 0$ must be a multiple eigenvalue with $AlgMult_{\lambda=0} > GeoMult_{\lambda=0}$, and this fact contradicts one of the previous results above revisited.

Then, from the subset of three pure behaviors, only the exponential one is possible in first-order reactions.

Case 4: Exponential-Oscillatory

This case appears when $\lambda \notin \mathbb{R}$ with $AlgMult_\lambda = GeoMult_\lambda$, and –at least theoretically– this situation may occur in the following FOCKM involving four chemical species: $A \overset{k_1}{\rightarrow} B$, $A \overset{k_2}{\rightarrow} C$, $A \overset{k_3}{\rightarrow} D$, $B \overset{k_4}{\rightarrow} C$, $C \overset{k_5}{\rightarrow} D$ and $D \overset{k_6}{\rightarrow} B$, where $k_1 = k_2 = k_3 = 3$, $k_4 = 2$ and $k_5 = k_6 = 1$.

The associated ODE system is:

$$\begin{pmatrix} [A]' \\ [B]' \\ [C]' \\ [D]' \end{pmatrix} = \begin{pmatrix} -9 & 0 & 0 & 0 \\ 3 & -2 & 0 & 1 \\ 3 & 2 & -1 & 0 \\ 3 & 0 & 1 & -1 \end{pmatrix} \begin{pmatrix} [A] \\ [B] \\ [C] \\ [D] \end{pmatrix} \quad (23)$$

The corresponding eigenvalues are: $\lambda_1 = -9$, $\lambda_{2,3} = -2 \pm i$ and $\lambda_4 = 0$, then, all the solutions are of the form:

$$C_1 e^{-9t} + C_2 e^{-2t} \cos t + C_3 e^{-2t} \sin t + C_4 \quad (24)$$

Consequently, the exponential-oscillatory behavior, at least theoretically, is possible in first-order reactions.

Case 5: Exponential-Polynomial

This case may happen in the consecutive reactions scheme: $A \overset{k_1}{\rightarrow} B \overset{k_2}{\rightarrow} C$, when $k_1 = k_2$.

In this situation, the ODE system is:

$$\begin{pmatrix} [A]' \\ [B]' \\ [C]' \end{pmatrix} = \begin{pmatrix} -k_1 & 0 & 0 \\ k_1 & -k_1 & 0 \\ 0 & k_1 & 0 \end{pmatrix} \begin{pmatrix} [A] \\ [B] \\ [C] \end{pmatrix} \quad (25)$$

The FOCKM matrix has a double eigenvalue $\lambda_1 = -k_1$, with associated eigenvectors multiples of $\overrightarrow{v_1} = (0, -1, 1)$, and there is a simple eigenvalue $\lambda_2 = 0$, being $\overrightarrow{v_2} = (0,0,1)$ the corresponding eigenvector.

Then, $AlgMult_{\lambda_1} = 2 > 1 = GeoMult_{\lambda_1}$, and so, the ODE system solutions are linear combinations of $e^{0t} = 1$, $e^{-k_1 t}$ and $te^{-k_1 t}$, being all the solutions for this system of the form:

$$C_1 + C_2 e^{-k_1 t} + C_3\, t\, e^{-k_1 t} \tag{26}$$

Then, at least from a theoretical viewpoint, exponential-polynomial curves are possible in FOCKM problems.

Case 6: Polynomial-Oscillatory

In order to obtain this type of solutions, it is necessary that $\lambda = 0 \pm ib$ with $b \neq 0$, eigenvalues of the FOCKM matrix, satisfy the inequality $AlgMult_{\lambda} > GeoMult_{\lambda}$. Independently of $AlgMult_{\lambda}$ and $GeoMult_{\lambda}$, this situation is impossible, since $Re(\lambda) = 0$ if and only if $\lambda = 0$.

Indeed, Figure 3, which is a graphical representation of a couple of eigenvalues $\lambda = 0 \pm ib$ with $b \neq 0$, shows the impossibility of this situation, since both eigenvalues lie outside the union of Gershgorin Circles. Then polynomial-oscillatory curves are not possible in FOCKM systems.

Case 7: Exponential-Polynomial-Oscillatory

In this case, let us consider a FOCKM involving six chemical substances, connected like in the following scheme: $A \xrightarrow{k_1} B$, $B \xrightarrow{k_2} C$, $C \xrightarrow{k_3} A$, $C \xrightarrow{k_4} B$, $D \xrightarrow{k_5} A$, $D \xrightarrow{k_6} E$, $E \xrightarrow{k_7} B$, $E \xrightarrow{k_8} F$, $F \xrightarrow{k_9} C$ and $F \xrightarrow{k_{10}} D$ where $k_1 = k_2 = \frac{5}{3}$, $k_3 = \frac{13}{15}$, $k_4 = \frac{4}{5}$ and $k_5 = k_6 = k_7 = k_8 = k_9 = k_{10} = 1$.

The associated ODE system is:

$$\begin{pmatrix} [A]' \\ [B]' \\ [C]' \\ [D]' \\ [E]' \\ [F]' \end{pmatrix} = \begin{pmatrix} -\frac{5}{3} & 0 & \frac{13}{15} & 1 & 0 & 0 \\ \frac{5}{3} & -\frac{5}{3} & \frac{4}{5} & 0 & 1 & 0 \\ 0 & \frac{5}{3} & -\frac{5}{3} & 0 & 0 & 1 \\ 0 & 0 & 0 & -2 & 0 & 1 \\ 0 & 0 & 0 & 1 & -2 & 0 \\ 0 & 0 & 0 & 0 & 1 & -2 \end{pmatrix} \begin{pmatrix} [A] \\ [B] \\ [C] \\ [D] \\ [E] \\ [F] \end{pmatrix} \tag{27}$$

The corresponding FOCKM matrix eigenvalues are:

$\lambda_1 = -1 \, , \lambda_2 = 0,$

$\lambda_{3,4} = -\frac{5}{2} - i\frac{\sqrt{3}}{2}$ and $\lambda_{5,6} = -\frac{5}{2} + i\frac{\sqrt{3}}{2}$.

The associated Jordan Canonical Form [27] is:

$$\mathbb{J} = \begin{pmatrix} -1 & 0 & 0 & 0 & 0 & 0 \\ 0 & 0 & 0 & 0 & 0 & 0 \\ 0 & 0 & \lambda_{3,4} & 1 & 0 & 0 \\ 0 & 0 & 0 & \lambda_{3,4} & 0 & 0 \\ 0 & 0 & 0 & 0 & \lambda_{5,6} & 1 \\ 0 & 0 & 0 & 0 & 0 & \lambda_{5,6} \end{pmatrix} \quad (28)$$

It is easy to observe that this Jordan matrix is not a diagonal form, and the non-zero entries in the superdiagonal confirm that the inequality $AlgMult_\lambda > GeoMult_\lambda$ is satisfied for the complex eigenvalues $\lambda_{3,4} = -\frac{5}{2} - i\frac{\sqrt{3}}{2}$ and $\lambda_{5,6} = -\frac{5}{2} + i\frac{\sqrt{3}}{2}$.

As a consequence of these facts, the solutions are linear combinations of:

$$e^{-t} \, , e^{0t} = 1 \, , \exp\left(-\tfrac{5}{2}t\right) \cos\left(\frac{\sqrt{3}}{2}t\right) , \exp\left(-\tfrac{5}{2}t\right) \sin\left(\frac{\sqrt{3}}{2}t\right),$$

and also:

$$t \, . \exp\left(-\tfrac{5}{2}t\right) \cos\left(\frac{\sqrt{3}}{2}t\right) \, , t \, . \exp\left(-\tfrac{5}{2}t\right) \sin\left(\frac{\sqrt{3}}{2}t\right).$$

Then, all the general solutions are of the form:

$$C_1 e^{-t} + C_2 + \exp\left(-\tfrac{5}{2}t\right)\left[C_3 \cos\left(\frac{\sqrt{3}}{2}t\right) + C_4 \sin\left(\frac{\sqrt{3}}{2}t\right)\right] +$$
$$+ \, t \, . \exp\left(-\tfrac{5}{2}t\right) \left[C_5 \cos\left(\frac{\sqrt{3}}{2}t\right) + C_6 \sin\left(\frac{\sqrt{3}}{2}t\right)\right] \quad (29)$$

Consequently, the exponential-polynomial-oscillatory behavior, at least in theory, is possible in cases where only first-order reactions are involved.

Conclusion

In this chapter, the possible concentration curves for mechanisms involving only first-order reactions were analyzed.

For this purpose, the corresponding ODE systems were studied from the viewpoint of their eigenvalues and eigenvectors, taking into account several previous theoretical results.

From this analysis, results that the only pure behavior that may occur is the exponential one.

On the contrary, the mixed behaviors (i.e., products of exponential, polynomial and/or trigonometric functions) are always possible, except the polynomial-oscillatory one.

In some cases, the possible behaviors were illustrated with real chemical reactions, whereas in others, only theoretical schemes were proposed.

Consequently, in the last cases, the problem of finding reactions with those possible behaviors remains open for further experimental research.

References

[1] Emanuel, N. M.; Kritsman, V. A.; Zaikov, G. E.; Markovi, N. *Chemical Kinetics and Chain Reactions: Historical Aspects*; Nova Science Publishers: New York; NY, 1995.

[2] Doggett, G.; Cockett, M.; Davies, A. G.; Phillips, D.; Derek-Woollins, J. *Maths for Chemists*; *2nd Ed.*; Royal Society of Chemistry: London, 2015.

[3] Steiner, E. *The Chemistry Maths Book*. Oxford University Press: Oxford, 2008.

[4] Zambelli, S. 2012. Chemical Kinetics, an Historical Introduction. In *Chemical Kinetics*, edited by Vivek Patel, 61-78. London: InTech Open Science.

[5] Wilhelmy, L., 1850. The Law of the Action of Acids upon Cane Sugar. *Annalen der Physik und Chemie* [Annals of Physics and Chemistry] 81, 413.

[6] Ebbing, D.; Gammon, S. D. *General Chemistry*. Cengage Learning: Boston, M A, 2016.

[7] Mortimer, R. G. *Physical Chemistry*. Academic Press: Burlington, M A. 2008.

[8] Teixeira-Dias, J. J. *Molecular Physical Chemistry*. Springer: Cham, 2017.

[9] Atkins, P.; De Paula, J. *Elements of Physical Chemistry*. Oxford University Press: Oxford, 2013.

[10] Agarwal, S. *Engineering Chemistry: Fundamentals and Applications, 2nd Ed.* Cambridge University Press: Cambridge, 2019.

[11] Hayes, R. E.; Mmbaga, J. P. *Introduction to Chemical Reactor Analysis*. CRC Press. Taylor & Francis Group: Boca Raton, FL, 2012.

[12] Tinoco, I.; Sauer; K.; Wang, J. C. *Physical Chemistry: Principles and Applications in Biological Sciences, 3rd Ed.* Prentice Hall: Hoboken, NJ, 1995.

[13] Martinez-Luaces, V. 2015. First-Order Chemical Kinetics Matrices and Stability of O.D.E. Systems. In *Advances in Linear Algebra Research*, edited by Ivan Kyrchei, 325-343, New York: Nova Science Publishers.

[14] Martinez-Luaces, V. 2017. Qualitative Behavior of Concentration Curves in First-Order Chemical Kinetics Mechanisms. In *Advances in Chemistry Research, Vol. 34*, edited by James C. Taylor, 139-169. New York: Nova Science Publishers.

[15] Martinez-Luaces, V. 2017. Matrices in Chemical Problems: Characterization, Properties and Consequences about the Stability of ODE Systems. In *Advances in Mathematics Research*, edited by Albert R. Baswell, 1-33, New York: Nova Science Publishers.

[16] Malik, B. A.; Benaissa, M. 2012. Glucose Chemistry. In *Dietary Sugars: Chemistry, Analysis, Function and Effects,* edited by Victor R. Preedy, V. R., 82-83, Royal Society of Chemistry: Cambridge.

[17] Martinez-Luaces, V. 2012. Chemical Kinetics and Inverse Modelling Problems. In *Chemical Kinetics*, edited by Vivek Patel, 61-78. London: InTech Open Science.

[18] Olah, G. A.; Goeppert, A.; Prakash, G. S. *Beyond Oil and Gas: the methanol economy*; Wiley-VCH: Weinheim, Germany, 2018.

[19] Fogler, H. S. *Essentials of Chemical Reaction Engineering*; Prentice Hall: Hoboken, NJ, 2011.

[20] Ricardo, H. J. *A Modern Introduction to Differential Equations*; Elsevier Science: Amsterdam, 2020.

[21] Zaitsev, V. F.; Polyanin, A. D. *Handbook of Exact Solutions for Ordinary Differential Equations*. CRC Press: Boca Raton, FL, 2002.

[22] Martinez-Luaces, V. 2016. Stability of ODE systems associated with first-order chemical kinetics mechanisms with and without final products. *Konuralp Journal of Mathematics*, 4(1), 80-87.

[23] Ahluwalia, V. K.; Dhingra, S; Gulati, A. *College Practical Chemistry*. Universities Press: Hyderabad, 2005.

[24] Kinoshita, S. *Pattern Formations and Oscillatory Phenomena*. Elsevier: Waltham, MA, 2013.

[25] Varga, R. S. *Geršgorin and His Circles*. Springer-Verlag: Berlin Heildelberg, 2011.

[26] Roman, S. *Advanced Linear Algebra*. Springer-Verlag: Berlin Heildelberg, 2013.

[27] Weintraub, S. H. *Jordan Canonical Form: Theory and Practice*. Springer Nature: Cham, 2022.

Chapter 3

New Kinetics of Free-Radical Nonbranched-Chain Addition with Competing Reactions

Michael M. Silaev*

Chemistry Faculty,
Lomonosov Moscow State University,
Vorobievy Gory, Moscow, Russia

Abstract

The kinetics of free-radical nonbranched-chain processes of addition to unsaturated compounds (such as alkenes, formaldehyde, dioxygen) was investigated. The aim of this study was the conclusion of simple kinetic equations to describe *ab initio* initiated nonbranched-chain processes of the saturated free-radical addition to the double bonds of unsaturated molecules in the binary reaction systems of saturated and unsaturated components. In the processes of this kind, the formation rate of the molecular addition products (1:1 adducts) as a function of concentration of the unsaturated component has a maximum. It is shown that a maximum in these curves arises from the competition between saturated and unsaturated components for reacting with the emerging 1:1 adduct radical.

Five reaction schemes are suggested for these addition processes. The proposed schemes include the reaction competing with chain propagation reactions through a reactive free radical. The chain evolution stage in these schemes involves three or four types of free radicals. One of them: $CH_2{=}C(CH_3)\dot{C}H_2$, $CH_2{=}CH\dot{C}HOH$, $H\dot{C}{=}O$, *o*-$CH_3C_6H_4CH_2O_4^{\bullet}$, or $HO_4^{\bullet}$ is relatively low-reactive and inhibits the chain process by shortening of the kinetic chain length. Based on the suggested schemes,

* Corresponding Author's Email: mmsilaev@gmail.com; michsil@yandex.ru.

In: A Closer Look at Chemical Kinetics
Editor: Victor Martinez-Luaces
ISBN: 979-8-88697-484-3

nine rate equations (containing one to three parameters to be determined directly) are deduced using quasi-steady-state treatment.

These equations provide good fits for the nonmonotonic (peaking) dependences of the formation rates of the molecular products (1:1 adducts) on the concentration of the unsaturated component in binary systems consisting of a saturated component (hydrocarbon, alcohol, etc.) and an unsaturated component (alkene, allyl alcohol, formaldehyde, or dioxygen). The unsaturated compound in these systems is both a reactant and an autoinhibitor generating low-reactive free radicals.

A similar kinetic description is applicable to the nonbranched-chain process of the free-radical hydrogen oxidation, in which the oxygen with the increase of its concentration begins to act as an oxidation autoinhibitor (or an antioxidant). The energetics of the key radical-molecule reactions is considered.

Keywords: binary system, unsaturated compound, low-reactive radical, autoinhibitor, competing reaction, nonbranched-chain addition, kinetic equation, rate, parameters, thermochemical data, energy

Introduction

A free radical may be low-reactive if its unpaired *p*-electron may be delocalized, e.g., over conjugated bonds as in the case of allyl radical $CH_2{=}CH\dot{C}H_2$ or along a double bond from carbon to the more electron-affine oxygen as in the case of formyl radical $H\dot{C}{=}O$. Note that the activity of a free radical is also connected to the reaction heat in which it participates. In nonbranched-chain processes of reactive free radical (addend) addition to double bonds of molecules, the formation of rather low-reactive free radicals in reactions, which are parallel to or competing with propagation *via* reactive radicals, lead to chain termination, because these low-reactive radicals do not participate in further chain propagation and because they decay when colliding with each other or with chain-carrier reactive radicals thus resulting in inefficient expenditure of the latter and process inhibition.

In similar processes involving the addend and inhibitor radicals in diffusion-controlled bimolecular chain-termination reactions of three types, the dependences of the rate of molecular 1:1 adduct formation on the concentration of the unsaturated component (which is the source of low-reactive free radicals in a binary system of saturated and unsaturated components) have a maximum, usually in the region of small (optimal) concentrations. The progressive inhibition of nonbranched chain processes upon exceeding this optimal concentration may be an element of self-regulation of the natural processes returning them to a steady-state condition.

Here, reactions of addition of reactive free radicals to multiple bonds of alkene, formaldehyde, and oxygen molecules to give 1:1 adduct radicals are taken as examples to consider the role of low-reactive free radicals as inhibitors of the nonbranched chain processes at moderate temperatures. In the case of oxidation, there are tetraoxyl 1:2 adduct radical arising upon the addition of a peroxyl 1:1 adduct radical to molecular oxygen at high enough concentrations of the latter.

The 1:1 adduct radical (which is the heaviest and the largest among the free radicals that result from the addition of one addend radical to the double bond of the molecule) may have an increased energy owing to the energy liberated in the transformation of a double bond into an ordinary bond (30–130 kJ mol^{-1} for the gas phase under standard conditions [1–4]). Therefore, it can decompose or react with one of the surrounding molecules in the place of its formation without diffusing in the solution and, hence, without participating in radical-radical chain termination reactions. Which of the two reactions of the adduct radical –the reaction with the saturated component or the reaction with the unsaturated component– dominates the kinetics of the process will depend on the reactivity and concentration ratios of the components in the binary system.

Earlier [5, 6], there were attempts to describe such peaking dependences fragmentarily, assuming that the saturated or unsaturated component is in excess, in terms of the direct and inverse proportionalities, respectively, that result from the simplification of a particular case of the kinetic equation set up by the quasi-steady-state treatment of binary copolymerization involving fairly long chains [5]. This specific equation is based on an irrational function, whose plot is a monotonic curve representing the dependence of the product formation rate on the concentration of the unsaturated component. This curve comes out of the origin of coordinates, is convex upward, and has an asymptote parallel to the abscissa axis. Replacing the component concentrations with the corresponding mole fractions generates a peak in this irrational function and thereby makes it suitable to describe the experimental data [7]. However, this circumstance cannot serve as a sufficient validation criterion for the mechanism examined, because the new property imparted to the function by the above artificial transformation does not follow from the solution of the set of algebraic equations that are set up for the reaction scheme accepted for the process in a closed system and express the equality of the steady-state formation and disappearance rates of the reactive intermediates.

This publication presents a comprehensive review of the nonbranched-chain kinetic models developed for particular types of additions of saturated free radicals to multiple bonds [8–14]. It covers free radical additions to alkenes [10, 11], their derivatives [8, 9], formaldehyde (first compound in the aldehyde homological series) [8, 9, 12], and molecular oxygen [13, 14] (which can add an unsaturated radical as well) yielding various 1:1 molecular adducts, whose formation rates as a function of the unsaturated compound concentration pass through a maximum (free radical chain additions to the C=N bond have not been studied adequately).

In the kinetic description of these non-telomerization chain processes, the reaction between the 1:1 adduct radical and the unsaturated molecule, which competes with chain propagation through a reactive free radical ($^{\bullet}PCl_2$, $C_2H_5\dot{C}HOH$, etc.), is included for the first time in the chain propagation stage. This reaction yields a low-reactive radical (such as $CH_2{=}C(CH_3)\dot{C}H_2$ or $H\dot{C}{=}O$) and thus leads to chain termination because this radical does not continue the chain and thereby inhibits the chain process [8]. We will consider kinetic variants for the case of comparable component concentrations with an excess of the saturated component [10, 11] and the case of an overwhelming excess of the saturated component over the unsaturated component [8, 9, 12]. Based on the reaction schemes suggested for the kinetic description of the addition process, we have derived kinetic equations with one to three parameters to be determined directly.

Reducing the number of unknown parameters in a kinetic equation will allow one to decrease the narrowness of the correlation of these parameters and avoid a sharp build-up of the statistical error in the nonlinear estimation of these parameters in the case of a limited number of experimental data points [15]. The rate constant of the addition of a free radical to the double bond of the unsaturated molecule, estimated as a kinetic parameter, can be compared to its reference value if the latter is known. This provides a clear criterion to validate the mathematical description against experimental data.

The kinetic equations were set up using the quasi-steady-state treatment. This method is the most suitable for processes that include eight to ten or more reactions and four to six different free radicals and are described by curves based on no more than three to seven experimental points. In order to reduce the exponent of the $2k_5[R_1^{\bullet}]^2$ term in the $d[R_1^{\bullet}]/dt = 0$ equation to unity [8], we used the following condition for the early stages of the process: $k_6 = \sqrt{2k_5 2k_7}$

[16] and, hence, $V_1 = V_5 + 2V_6 + V_7 = (\sqrt{2k_5}\,[R_1^{\bullet}] + \sqrt{2k_7}\,[R_2^{\bullet}])^2$. Here, $[R_1^{\bullet}]$ and $[R_2^{\bullet}]$ are the concentrations of the addend radical and the low-reactive (inhibitor) radical, respectively; V_1 is the initiation rate; V_5, $2V_6$, and V_7 are the rates of the three types of diffusion-controlled quadratic-law chain termination reactions; $2k_5$ and $2k_7$ are the rate constants of the loss of identical free radicals *via* the reactions $R_1^{\bullet} + R_1^{\bullet}$ and $R_2^{\bullet} + R_2^{\bullet}$, respectively; k_6 is the rate constant of the loss of different free radicals via the $R_1^{\bullet} + R_2^{\bullet}$ reaction (see *Schemes 1–5*). The kinetic equations thus obtained fit the peaking rate curves well throughout the range of unsaturated component concentrations in the binary systems. Our mathematical simulation was based on experimental data obtained for γ-radiation-induced addition reactions for which the initiation rate V_1 is known.

The analysis of stable liquid-phase products was carried out by the gas chromatographic method.

Addition to the C=C Bond of Alkenes and Their Derivatives

When reacting with alkenes not inclined to free-radical polymerization, the free radicals originating from inefficient saturated telogens, such as alcohols [17] and amines [18], usually add to the least substituted carbon atom at the double bond, primarily yielding a free 1:1 adduct radical. This radical accumulates an energy of 90–130 kJ mol^{-1}, which is released upon the transformation of the C=C bond to an ordinary bond (according to the data reported for the addition of nonbranched C_1–C_4 alkyl radicals to propene and of similar C_1 and C_2 radicals to 1-butene in the gas phase under standard conditions [1–4]). Such adduct radicals, which do not decompose readily for structural reasons, can abstract the most labile atom from a neighbor molecule of the saturated or unsaturated component of the binary reaction system, thus turning into a 1:1 adduct molecule.

The consecutive and parallel reactions involved in this free-radical nonbranched-chain addition process are presented below (*Scheme 1*). In the case of comparable component concentrations with a non-overwhelming excess of the saturated component, extra reaction (1b) ($k_{1b} \neq 0$) is included in the initiation stage [10, 11]. In the case of an overwhelming excess of the saturated component reaction (1b) is ignored ($k_{1b} = 0$) [8, 9, 12].

Comparable Component Concentrations

Scheme 1

Chain initiation:

1. $I \xrightarrow{2k_1} 2R_0^{\bullet}$;

1a. $R_0^{\bullet} + R_1A \xrightarrow{k_{1a}} R_0A + R_1^{\bullet}$;

1b. $R_0^{\bullet} + R_2A \xrightarrow{k_{1b}} R_0A + R_2^{\bullet}$.

Chain propagation:

2. $R_1^{\bullet} + R_2B \xrightarrow{k_2} R_3^{\bullet}$;

3. $R_3^{\bullet} + R_1A \xrightarrow{k_3} R_3A + R_1^{\bullet}$.

Inhibition:

4. $R_3^{\bullet} + R_2B \xrightarrow{k_4} R_3B + R_2^{\bullet}$.

Chain termination:

5. $2R_1^{\bullet} \xrightarrow{2k_5}$ Prod ;

6. $R_1^{\bullet} + R_2^{\bullet} \xrightarrow{k_6}$ Prod ;

7. $2R_2^{\bullet} \xrightarrow{2k_7}$ Prod.

In this scheme, I is an initiator (e.g., a peroxide [5, 12, 13]); $R_0^{\bullet}$ is a reactive (initiating) radical; A and B are hydrogen or halogen atoms [2, 5, 17–24]; $R_1^{\bullet}$ is $^{\bullet}PCl_2$ [19], $^{\bullet}CCl_3$ [20], alkyl [2, 5], 1-hydroxyalkyl [5, 6, 17, 22–24], or a similar functionalized reactive addend radical [5]; $R_2^{\bullet}$ is an alkenyl radical (allyl or higher) [2, 5, 17–22], 1-hydroxyalkenyl [5, 17, 18, 23, 24], or a similar functionalized low-reactive radical (inhibitor) [5, 18]; $R_3^{\bullet}$ is a saturated reactive 1:1 adduct radical; R_0A, R_0B, and R_1A are saturated molecules; R_2B is an unsaturated molecule (alkene or its derivative); R_3A and R_3B are 1:1

adduct molecules; Prod designates the molecular products resulting from the dimerization or disproportionation of free radicals. The chain evolution (propagation and inhibition) stage of *Scheme 1* includes consecutive reactions 2 and 3, parallel (competing) reaction pairs 3 and 4, and consecutive-parallel reaction pair 2–4.

The initiation reaction 1 is either the decomposition of a chemical initiator [5, 17, 18] or a reaction induced by light [5, 17, 18] or ionizing radiation [19–23]. The overall rate of chain initiation (reactions 1, 1a, and 1b) is determined by the rate of the rate-limiting step ($k_{1b} > k_{1a}$). The reaction between the free radical $R_2^\bullet$, which results from reactions 1b and 4, and the saturated molecule R_1A is energetically unfavorable because it implies the formation of the free radical $R_1^\bullet$, which is less stable than the initial one. The addition reaction 2 may be accompanied by the parallel abstraction reaction.

$$\text{2a.}\quad R_1^\bullet + R_2B \xrightarrow{k_{2a}} R_1B + R_2^\bullet$$

which yields the product R_1B *via* a nonchain mechanism. Reaction 2a does not regenerate the addend radical $R_1^\bullet$ and is not necessary for a kinetic description of the process, because the rate ratio of reactions 2 and 2a, $V_2/V_{2a} = k_2/k_{2a}$, is independent of the concentration of the unsaturated component R_2B in the system. The inhibition of the nonbranched-chain addition process is due to reaction 4, in which the adduct radical $R_3^\bullet$ is spent in an inefficient way, since this reaction, unlike reaction 3, does not regenerate $R_1^\bullet$. The inhibiting effect is also due to the loss of chain carriers $R_1^\bullet$ through their collisions with low-reactive unsaturated radicals $R_2^\bullet$, but to a much lesser extent.

The rates of the formation (V, mol dm^{-3} s^{-1}) of the 1:1 adducts R_3A (*via* a chain mechanism) and R_3B (*via* a nonchain mechanism) in reactions 3 and 4 are given by the equations:

$$V_3(R_3A) = \frac{[\gamma l/(\gamma l + x)]V_1 \alpha l k_2 x}{k_2 x^2 + (\alpha l + x)\sqrt{2k_5 V_1}}, \tag{1}$$

$$V_4(R_3B) = \frac{[\gamma l/(\gamma l + x)]V_1 k_2 x^2}{k_2 x^2 + (\alpha l + x)\sqrt{2k_5 V_1}}, \tag{2}$$

where V_1 is the rate of the initiation reaction 1; $l = [R_1A]$ and $x = [R_2B]$ are the molar concentrations of the initial components, with $l > x$; k_2 is the rate constant of the addition of the $R_1^{\bullet}$ radical from the saturated component R_1A to the unsaturated molecule R_2B (reaction 2), and $\gamma = k_{1a}/k_{1b}$ and $\alpha = k_3/k_4$ are the rate constant ratios for competing (parallel) reactions (α is the first chain-transfer constant for the free-radical telomerization process [5]). The rate ratio for the competing reactions is $V_3/V_4 = \alpha l/x$, and the chain length is $\nu = V_3/V_1$.

Earlier mathematical simulation [8] demonstrated that replacing the adduct radical R_3 with the radical R_2 [5] in the reaction between identical radicals and in the reaction involving R_1 gives rise to a peak in the curve of the 1:1 adduct formation rate as a function of the concentration of the unsaturated component. Reaction 1b, which competes with reaction 1a, is responsible for the maximum in the curve described by Equation (2), and reaction 4, which is in competes with reaction (3), is responsible for the maximum in the curve defined by Equation (1).

The number of unknown kinetic parameters to be determined directly (k_2, α, and γ) can be reduced by introducing the condition $\gamma \cong \alpha$, which is suggested by the chemical analogy between the competing reaction pairs 1a–1b and 3–4. For example, the ratios of the rate constants of the reactions of $^{\bullet}OH$, $CH_3O^{\bullet}$, $^{\bullet}CH_3$, $NO_3^{\bullet}$, and $H_2PO_4^{\bullet}$ with methanol to the rate constants of the reactions of the same radicals with ethanol in aqueous solution at room temperature are 0.4–0.5 [25, 26]. For the same purpose, the rate constant of reaction 2 in the kinetic equation can be replaced with its analytical expression:

$$k_2 = \alpha l_m \sqrt{2k_5 V_1} / x_m^2 ,$$

which is obtained by solving the quadratic equation following from the reaction rate extremum condition $\partial V_{3,4}(1{:}1\ \text{Adduct})/\partial x = 0$, where $V_{3,4}(1{:}1\ \text{Adduct}) = V_3 + V_4$. After these transformations, the overall formation rate equation for the 1:1 adducts R_3A and R_3B (which may be identical, as in the case of R_3H [5, 8, 9, 12, 13, 18–21]), appears as

$$V_{3,4}(1{:}1\ \text{Adduct}) = \frac{V_1 \alpha l k_2 x}{k_2 x^2 + (\alpha l + x)\sqrt{2k_5 V_1}} \quad (3)$$

$$= \frac{V_1 \alpha l x}{x^2 + (\alpha l + x) x_m^2 / \alpha l_m}, \tag{3a}$$

where l_m and x_m are the component concentrations l and x at the points of maximum of the function. Provided that V_1 is known, the only parameter in Equation (3a) to be determined directly is α. If V_1 is known only for the saturated component R_1A, then, for the binary system containing comparable R_1A and R_2B concentrations, it is better to use the quantity λV_1, where $\lambda = l/(l + x)$ is the mole fraction of R_1A, in place of V_1 in Equations (3) and (3a).

The two variable concentrations in the kinetic equation (3) – l and x – can be reduced to one variable by replacing them with the corresponding mole fractions. Substituting the expression:

$$k_2 = \left\{\alpha\left[(1/\chi_m) - 1\right]^2 - 1\right\}\sqrt{2k_5V_1} \Big/ (l_m + x_m),$$

derived from the rate extremum condition, into this transformed equation for the binary system containing comparable component concentrations, we obtain:

$$V_{3,\,4}(\text{Adduct 1:1}) = \frac{V_1 \alpha (1-\chi)\chi}{\chi^2 + \left[\alpha(1-\chi) + \chi\right] \Big/ \left\{\alpha\left[(1/\chi_m) - 1\right]^2 - 1\right\}}, \tag{3b}$$

where $1 - \chi = l/(l + x)$ and $\chi = x/(l + x)$ are the mole fractions of the components R_1A and R_2B ($0 < \chi < 1$), respectively, and χ_m is the χ value at the point of maximum.

The overall formation rate of the 1:1 adducts R_3A and R_3B is a sophisticated function of the formation and disappearance rates of the radicals $R_1^\bullet$ and $R_2^\bullet$: $V(R_3A, R_3B) = (V_{1a} + V_3 - V_5) - (V_{1b} + V_4 - V_7)$.

The application of the above rate equations to particular single nonbranched-chain additions is illustrated in Figure 1. Curve *1* represents the results of simulation in terms of Equation (3b) for the observed 1:1 adduct formation rate as a function of the mole fraction of the unsaturated component in the phosphorus trichloride–methylpropene[1] reaction system at 303 K [19].

---

[1] In an earlier work [10], the methylpropene concentration in this system was overvalued by a factor of 1.7 when it was derived from the mole fractions given in [19].

In this simulation, the ^{60}Co γ-radiation dose rate was set at $P = 0.01$ Gy s^{-1} and the initiation yield was taken to be $G(^{\cdot}PCl_2) = 2.8$ particles per 100 eV (1.60×10^{-17} J) of the energy absorbed by the solution [19]. The product of reaction 3 is $Cl_2PCH_2C(Cl)(CH_3)CH_3$ (two isomers), $V_1 = 4.65 \times 10^{-9}$ mol dm^{-3} s^{-1} at $\chi = 0$, and $2k_5 = 3.2 \times 10^8$ dm^3 mol^{-1} s^{-1}. This leads to $\alpha = (2.5 \pm 0.4) \times 10^3$, and the rate constant of reaction 2 derived from this α value is $k_2 = (1.1 \pm 0.2) \times 10^4$ dm^3 mol^{-1} s^{-1}.

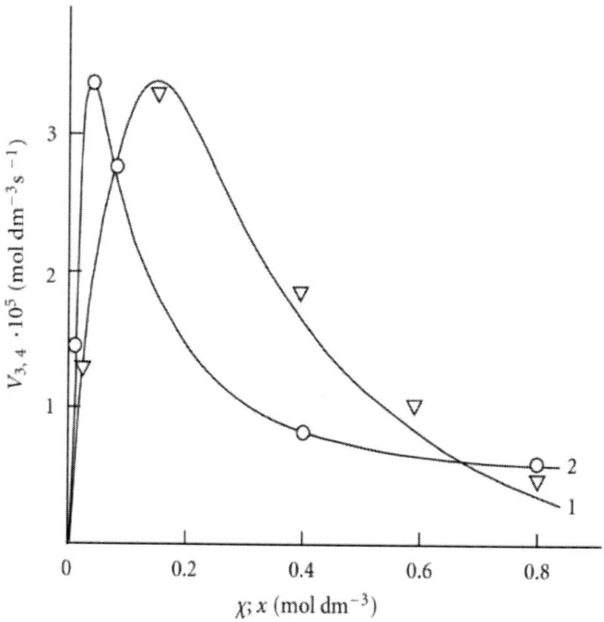

Figure 1. Reconstruction of the functional dependences (curves) of the product formation rates $V_{3,4}$ (*1*, ▽) on the mole fraction of the unsaturated component (χ) from empirical data (symbols) using Equation (3b) (model optimization with respect to the parameter α) for the phosphorus trichloride–methylpropene reaction system at 303 K [19] (standard deviation of $S_Y = 2.58 \times 10^{-6}$) and (*2*, ○) on the concentration of the unsaturated component (x) from empirical data (symbols) using Equation (4a) (model optimization with respect to V_1, x_m, and α) for the 2-propanol–2-propen-1-ol system at 433 K [23] ($S_Y = 5.91 \times 10^{-7}$).

Note that, if the R_2–B bond dissociation energy for the unsaturated component of the binary system is approximately equal to or above, not below, the R_1–A bond dissociation energy for the saturated component, then the rate of reaction 4 relative to the rate of the parallel reaction 3 (chain propagation through the reactive free radical $R_1^{\bullet}$) will be sufficiently high for an adequate description of R_3A and R_3B adduct formation in terms of Equations (1)–(3b) only at high temperatures [20]. In the phosphorus trichloride–propene system, the difference between the R_2–B (B = H) and R_1–A (A = Hal) bond dissociation energies in the gas phase under standard conditions [1] is as small as 5 kJ mol^{-1}, while in the tetrachloromethane–methylpropene (or cyclohexene) and bromoethane–2-methyl-2-butene systems, this difference is 20.9 (37.7) and ~24 kJ mol^{-1}, respectively.

Excess of the Saturated Component

If the concentration of the saturated component exceeds the concentration of the unsaturated component in the binary system, reaction 1b can be neglected. If this is the case (k_{1b} = 0), then, in the numerators of the rate equations for reactions 3 and 4 (Equations (1) and (2)), $\gamma l/(\gamma l + x) = 1$ and the overall rate equation for the formation of the 1:1 adducts R_3A and R_3B will appear as:

$$V_{3,\,4}(\text{1:1 Addduct}) = \frac{V_1(\alpha l + x)k_2 x}{k_2 x^2 + (\alpha l + x)\sqrt{2k_5 V_1}} \tag{4}$$

$$= \frac{V_1 x}{\dfrac{x^2}{\alpha l + x} + \left(\dfrac{\sqrt{\alpha l_m}}{x_m} + \dfrac{1}{\sqrt{\alpha l_m}}\right)^{-2}}, \tag{4a}$$

where the parameters are designated in the same way as in Equations (1)–(3a), $l >> x$,

$$k_2 = [(\sqrt{\alpha l_m}/x_m) + (1/\sqrt{\alpha l_m})]^2 \sqrt{2k_5 V_1}$$

is determined from the condition $\partial V_{3,4}(\text{1:1 Adduct})/\partial x = 0$.

The rate equations for the chain termination reactions 5–7 (*Scheme 1*, k_{1b} = 0) are identical to Equations (12)–(14) (see below) with $\beta = 0$.

Note that, if it is necessary to supplement *Scheme 1* for k_{1b} = 0 with the formation of R_1B *via* the possible nonchain reaction 2a (which is considered in Section 2.1), the parameter k_{2a} should be included in the denominator of Equation (4) to obtain:

$$k_2x^2 + (\alpha l + x)(k_{2a}x + \sqrt{2k_5V_1}).$$

The analytical expression for k_2 in the case of $k_{2a} \neq 0$ is identical to the expression for k_2 for Equation (4). The equation for the rate $V_{2a}(R_1B)$ can be derived by replacing k_2 with k_{2a} in the numerator of Equation (4) containing k_{2a} in its denominator.

Curve *2* in Figure 1 illustrates the good fit between Equation (4a) and the observed 1:1 adduct formation rate as a function of the concentration of the unsaturated component in the reaction system 2-propanol–2-propen-1-ol at 433 K [8, 9]. In this description, we used a ^{60}Co γ-radiation dose rate of $P = 4.47$ Gy s^{-1} [23].

The product of reactions 3 and 4 is $CH_3(CH_3)C(OH)CH_2CH_2CH_2OH$, and $2k_5 = 1.0 \times 10^{10}$ dm^3 mol^{-1} s^{-1}.

The following parameters were obtained: $V_1 = (3.18 \pm 0.4) \times 10^6$ mol dm^{-3} s^{-1}, $x_m = (3.9 \pm 0.5) \times 10^{-2}$ mol dm^{-3}, and $\alpha = (6.8 \pm 0.8) \times 10^{-2}$. The rate constant of reaction 2 derived from this α is $k_2 = (1.0 \pm 0.14) \times 10^5$ dm^3 mol^{-1} s^{-1}.

Addition to the C=O Bond of Formaldehyde

Free radicals add to the carbon atom at the double bond of the carbonyl group of dissolved free (unsolvated, monomer) formaldehyde. The concentration of free formaldehyde in the solution at room temperature is a fraction of a percent of the total formaldehyde concentration, which includes formaldehyde chemically bound to the solvent [27]. The concentration of free formaldehyde exponentially increases with increasing temperature [28]. The energy released as a result of this addition, when the C=O bond is converted into an ordinary bond, is 30 to 60 kJ mol^{-1} (according to the data on the addition of C_1–C_4 alkyl radicals in the gas phase under standard conditions [1–4]). The resulting free 1:1 adduct radicals can both abstract hydrogen atoms from the nearest-neighbor molecules of the solvent or unsolvated formaldehyde and, due to its structure, decompose by a monomolecular mechanism including isomerization [9, 12].

Addition of Free 1-Hydroxyalklyl Radicals with Two or More Carbon Atoms

Free 1-hydroxyalkyl radicals (which result from the abstraction of a hydrogen atom from the carbon atom bonded to the hydroxyl group in molecules of saturated aliphatic alcohols but methanol under the action of chemical initiators [29, 30], light [17, 31], or ionizing radiation [32, 33]) add at the double bond of free formaldehyde dissolved in the alcohol, forming 1,2-alkanediols [8, 9, 12, 29–36], carbonyl compounds, and methanol [8, 33] *via* the chaining mechanism. (The yields of the latter two products in the temperature range of 303 to 448 K are one order of magnitude lower.)

In these processes, the determining role in the reactivity of the alcohols can be played by the desolvation of formaldehyde in alcohol–formaldehyde solutions, which depends both on the temperature and on the polarity of the solvent [28, 33]. For the γ-radiolysis of 1(or 2)-propanol–formaldehyde system at a constant temperature, the dependences of the radiation-chemical yields of 1,2-alkanediols and carbonyl compounds as a function of the formaldehyde concentration show maxima and are symbatic [8, 32]. For a constant total formaldehyde concentration of 1 mol dm^{-3}, the dependence of the 1,2-alkanediol yields as a function of temperature for 303–473 K shows a maximum, whereas the yields of carbonyl compounds and methanol increase monotonically [33] (along with the concentration of free formaldehyde [28]).

In addition to the above products, the nonchain mechanism in the γ-radiolysis of the solutions of formaldehyde in ethanol and 1- and 2-propanol gives ethanediol, carbon monoxide, and hydrogen in low radiation-chemical yields (which, however, exceed the yields of the same products in the γ-radiolysis of individual alcohols) [8, 9, 33]. The available experimental data can be described in terms of the following scheme of reactions:

Scheme 2

Chain initiation:

$$1.\ \mathrm{I} \xrightarrow{2k_1} 2\,\mathrm{R}_0^{\bullet};$$

$$1\mathrm{a}.\ \mathrm{R}_0^{\bullet} + \mathrm{ROH} \xrightarrow{k_{1\mathrm{a}}} \mathrm{ROH} + {}^{\bullet}\mathrm{R}_{(-\mathrm{H})}\mathrm{OH}.$$

Chain propagation:

$$2.\ {}^{\bullet}R_{(-H)}OH + CH_2O \xrightarrow{k_2} R_{(-H)}(OH)CH_2O^{\bullet};$$

$$3.\ R_{(-H)}(OH)CH_2O^{\bullet} + ROH \xrightarrow{k_3} R_{(-H)}(OH)CH_2OH + {}^{\bullet}R_{(-H)}OH;$$

$$3a.\ R_{(-H)}(OH)CH_2O^{\bullet} \xrightarrow{k_{3a}} R_{(-2H)}HO + {}^{\bullet}CH_2OH \quad (\text{or } R'R''CO + {}^{\bullet}CH_2OH);$$

$$3b.\ {}^{\bullet}CH_2OH + ROH \xrightarrow{k_{3b}} CH_3OH + {}^{\bullet}R_{(-H)}OH.$$

Inhibition:

$$4.\ R_{(-H)}(OH)CH_2O^{\bullet} + CH_2O \xrightarrow{k_4} R_{(-H)}(OH)CH_2OH + {}^{\bullet}CHO.$$

Chain termination:

$$5.\ 2{}^{\bullet}R_{(-H)}OH \xrightarrow{2k_5} R_{(-H)}(OH)R_{(-H)}OH \quad (\text{or: } ROH + R_{(-2H)}HO,\ ROH + R'R''CO);$$

$$6.\ {}^{\bullet}R_{(-H)}OH + {}^{\bullet}CHO \xrightarrow{k_6} R_{(-H)}(OH)CHO \quad (\text{or: } R_{(-2H)}HO + CH_2O,\ R'R''CO + CH_2O,\ ROH + CO);$$

$$7.\ 2{}^{\bullet}CHO \xrightarrow{2k_7} HC(O)CHO \quad (\text{or: } CH_2O + CO,\ 2CO + H_2).$$

In these reactions, I is an initiator, e.g., a peroxide [29, 30]; $R_0^{\bullet}$, some reactive radical (initiator radical); R, an alkyl; ROH, a saturated aliphatic alcohol, either primary or secondary, beginning from ethanol; CH_2O, the unsaturated molecule – free formaldehyde; ${}^{\bullet}CH_2OH$, the 1-hydroxymetyl fragment radical; ${}^{\bullet}R_{(-H)}OH$, the reactive 1-hydroxyalkyl addend radical, beginning from 1-hydroxyethyl; $R_{(-H)}(OH)CH_2O^{\bullet}$, the reactive hydroxy-alkoxyl 1:1 adduct radical; ${}^{\bullet}CHO$, the low-reactive formyl radical (inhibitor radical); R_0H, the molecular product; $R_{(-H)}(OH)CH_2OH$, 1,2-alkanediol; $R_{(-2H)}HO$, an aldehyde in the case of a primary alcohol and an R'R"CO ketone in the case of a secondary alcohol; $R_{(-H)}(OH)R_{(-H)}OH$, a vicinal alkanediol; $R_{(-H)}(OH)CHO$, a hydroxyaldehyde. The chain evolution stage of *Scheme 2* includes consecutive reaction pairs 2–3, 2–3a, and 3a–3b; parallel (competing) reaction pairs 3–3a, 3–3b, 3–4, and 3a–4; and consecutive–parallel reactions 2 and 4.

Scheme 2 does not include the same types of radical-molecule reactions as were considered in Section 2.1 for *Scheme 1*. In addition, it seems unlikely that free adduct radicals will add to formaldehyde at higher temperatures the reaction of adding is unlikely because this would result in an ether bond. The addition of hydroxymethyl radicals to formaldehyde, which competes with reaction 3b, is not included as well, because there is no chain formation of ethanediol at 303–448 K [33]. At the same time, small amounts of ethanediol can form *via* the dimerization of a small fraction of hydroxymethyl radicals, but this cannot have any appreciable effect on the overall process kinetics. The addition of free formyl radicals to formaldehyde cannot proceed at a significant rate, as is indicated by the fact that there is no chain formation of glycol aldehyde in the systems examined [33].

The mechanism of the decomposition of the free adduct radical *via* reaction 3a, which includes the formation of an intramolecular H···O bond and isomerization, can be represented as follows [8, 9, 12]:

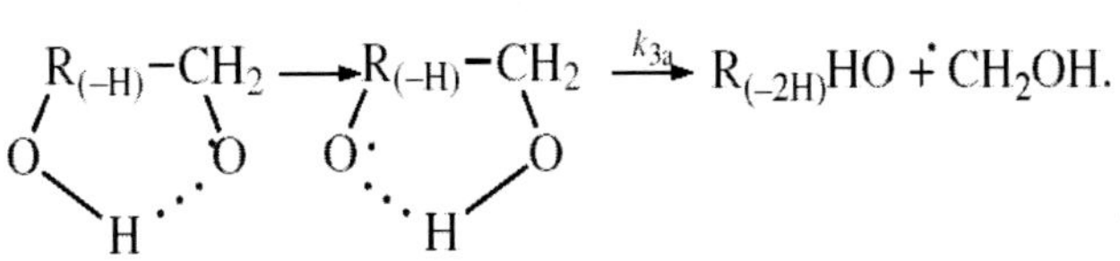

The probability of the occurrence of reaction 3a should increase with increasing temperature. This is indicated by experimental data presented above [8, 9, 12]. The decomposition of the hydroxyalkoxyl radical. $R_{(-H)}(OH)CH_2O^{\bullet}$ (reaction 3a) is likely endothermic. The endothermic nature of reaction 3a is indirectly indicated by the fact that the decomposition of simple C_2–C_4 alkoxyl radicals $RO^{\bullet}$ in the gas phase is accompanied by heat absorption: (ΔH°_{298}= 30–90 kJ mol^{-1} [2–4]). Reaction 3b, subsequent to reaction 3a, is exothermic, and its heat for C_2–C_3 alcohols in the gas phase is ΔH°_{298} = −40 to −60 kJ mol^{-1} [2–4]. As follows from the above scheme of the process, reactions 3a and 3b, in which the formation and consumption of the highly reactive free radical hydroxymethyl take place (at equal rates under steady-state conditions), can be represented as a single bimolecular reaction 3a,b occurring in a "cage" of solvent molecules.

The free formyl radical resulting from reaction 4, which competes with reactions 3 and 3a, is comparatively low-reactive because its spin density can be partially delocalized from the carbon atom *via* the double bond toward the oxygen atom, which possesses a higher electron affinity [1]. For example, in contrast to the methyl and alkoxyl π-radicals, the formyl σ-radical can be stabilized in glassy alcohols at 77 K [37]. In the gas phase, the dissociation energy of the C–H bond in formyl radicals is half that for acetyl radicals and is about 5 times lower than the dissociation energy of the C_α–H bond in saturated C_1–C_3 alcohols [1].

As distinct from reactions 3 and 3a,b, reaction 4 leads to an inefficient consumption of hydroxyalkoxyl adduct radicals, without regenerating the initial 1-hydroxyalkyl addend radicals. Reaction 4 together with reaction 6 (mutual annihilation of free formyl and chain-carrier 1-hydroxyalkyl radicals) cause the inhibition of the nonbranched-chain process. For the disproportionation of the free radicals, the heats of reactions 5–7 for C_1–C_3 alcohols in the gas phase vary in the range of $\Delta H^\circ_{298} = -135$ to -385 kJ mol^{-1} [1–4].

The rates of the chain formation of 1,2-alkanediols in reaction 3 (and their nonchain formation in reaction 4), carbonyl compounds in reaction 3a, and methanol in reaction 3b are given by the following equations:

$$V_{3,4}(R_{(-H)}(OH)CH_2OH) = \frac{V_1(\alpha l + x)k_2 x}{k_2 x^2 + (\alpha l + \beta + x)\sqrt{2k_5 V_1}}, \quad (5)$$

$$V_{3a}(R_{(-2H)}HO) = V_{3b}(CH_3OH) = \frac{V_1 \beta k_2 x}{k_2 x^2 + (\alpha l + \beta + x)\sqrt{2k_5 V_1}}, \quad (6)$$

where V_1 is the initiation rate, l is the molar concentration of the saturated alcohol at a given total concentration of formaldehyde[2] dissolved in it, x is the molar concentration of free formaldehyde ($l >> x$), k_2 is the rate constant of reaction 2 (addition of 1-hydroxyalkyl free radical to free formaldehyde), and $\alpha = k_3/k_4$ and $\beta = k_{3a}/k_4$ (mol dm^{-3}) are the ratios of the rate constants of the competing (parallel) reactions. Estimates of $2k_5$ were reported by Silaev *et al.*

---

[2] The alcohol concentration in alcohol–formaldehyde solutions at any temperature can be estimated by the method suggested in [38, 39]. The data necessary for estimating the concentration of free formaldehyde using the total formaldehyde concentration in the solution are reported by Silaev *et al.* [28, 39].

[39, 40]. From the extremum condition for the reaction 3a rate function, $\partial V_{3a}/\partial x = 0$, we derived the following analytical expression:

$$k_2 = (\alpha l_m + \beta)\sqrt{2k_5V_1}\,/\,x_m^2\,.$$

The overall process rate is a complicated function of the formation and disappearance rates of the $^{\bullet}R_{(-H)}OH$ and $^{\bullet}CHO$ free radicals: $V(R_{(-H)}(OH)CH_2OH, R_{(-2H)}HO, CH_3OH) = V_{1a} + V_3 + V_{3b} - V_4 - V_5 + V_7$. The ratios of the rates of the competing reactions are $V_3/V_4 = \alpha l/x$ and $V_{3a}/V_4 = \beta/x$, and the chain length is $\nu = (V_3 + V_{3a})/V_1$. The ratio of the rates of formation of 1,2-alkanediol and the carbonyl compound is a simple linear function of x: $V_{3,4}(R_{(-H)}(OH)CH_2OH)/V_{3a}(R_{(-2H)}HO) = (k_4/k_{3a})x + (k_3/k_{3a})l$. The equations for the rates of chain-termination reactions 5–7 are identical to Equations (12)-(14) (see below, Section 4.1).

Neutral formaldehyde solutions in alcohols at room temperature primarily consist of a mixture of formaldehyde polymer solvates reversibly bound to alcohols; these polymer solvates differ in molecular mass and have the general formula $RO(CH_2O)_nH$, where n = 1–4 [27]. The concentration of formaldehyde that occurs in solution as a free, unsolvated active species chemically unbound to the solvent (this species is capable of scavenging free radicals) at room temperature is lower than a percent of the total formaldehyde concentration [27]. The concentration x of the free formaldehyde species in solutions was determined by high-temperature UV spectrophotometry in the range 335–438 K at the total formaldehyde concentration c_0 (free and bound species including the concentration of polymer solvates) of 1.0–8.4 mol dm^{-3} in water, ethanediol, methanol, ethanol, 1-propanol, 2-propanol, and 2-methyl-2-propanol [28] (see Table of the Appendix). This concentration increases with temperature according to an exponential law, and it can be as high as a few percent of the total concentration in solution under the test conditions, up to 19.3% in the case of 2-methyl-2-propanol at a total concentration of 1.0 mol dm^{-3} and a temperature of 398 K. The following empirical equation relating the concentration x (mol dm^{-3}) of free formaldehyde to temperature T (K) and the total concentration c_0 in the solution (measured at room temperature), was developed by the treatment of 101 data points [28, 39]:

$$\lg x = -a\,(10^3/T) + b + h \lg c_0, \qquad (7)$$

where the coefficients a and b were calculated as the parameters of a straight-line equation by the least-squares technique from the dependence of lg x on $1/T$ at c_0 = 1.0 mol dm^{-3} for various solvents, and the coefficient h was obtained as the average value of the slopes of lg x as linear functions of lg c_0 at various series of fixed temperatures. Table 1 summarizes these coefficients for each solvent. As regards the experimental data, the error in the calculations of the concentration x of free formaldehyde made by Equation (7) in the specified temperature range was no higher than 25%.

Table 1. Coefficients of the empirical Equation (7) for the estimation of the concentration x of free formaldehyde in polar solvent–formaldehyde systems

Solvent	Coefficient		
	a	b	h
Water	2.36	4.45	0.80
Ethanediol	1.83	2.60	1.28
Methanol	3.11	5.58	$0.22\ c_0 / \lg c_0$
Ethanol	3.10	5.92	$1.10\ (10^3/T) - 1.44$
1-Propanol	2.42	4.47	1.30
2-Propanol	2.42	4.64	1.05
2-Methyl-2-propanol	3.19	7.31	0.96

On the assumption that the dependence of the density of a given solution on the concentration of formaldehyde is similar to the analogous linear dependence found for aqueous formaldehyde solutions (0–14 mol dm^{-3}; 291 K) [27], the concentrations l_T (mol dm^{-3}) of alcohols in alcohol–formaldehyde solutions at a certain temperature can be estimated by the equation:

$$l_{\mathrm{T}} = \frac{(10^3 d - 21.6\, c_0) d_{\mathrm{T}}}{(d + 8.4 \times 10^{-3}\, c_0) M}, \tag{8}$$

where c_0 is the total formaldehyde concentration (mol dm^{-3}); M is the molecular mass (g mol^{-1}) of the solvent; d and d_T are the solvent densities (g cm^{-3}) at room and given temperatures, respectively; the coefficients 8.4 × 10^{-3} and 21.6 have the units of 10^3 g mol^{-1} and g mol^{-1}, respectively [38].

Earlier [28], it was found that the concentration x of the free formaldehyde species decreased with the solvent permittivity D_{298} at a constant temperature. Water is an exception. Although water is more polar than alcohols, the concentration x of free formaldehyde in an aqueous solution is anomalously high and reaches the level of its concentration in 2-propanol, all other factors being the same (see Figure 2) [28, 39].

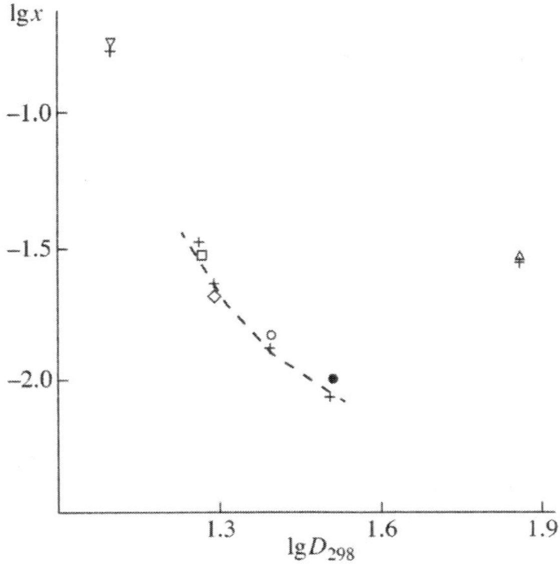

Figure 2. Logarithmic plot of the experimental concentrations x (mol dm^{-3}) of free formaldehyde at its total concentration c_0 = 1.0 mol dm^{-3} and 395 ± 3 K in (△) water, (●) methanol, (○) ethanol, (◇) 1-propanol, (□) 2-propanol, and (▽) 2-methyl-2-propanol as functions of the permittivity D_{298} of these solvents (+ refers to the concentrations x in the above solvents calculated by empirical Equation (8)).

This can be due to the specific instability of hydrated formaldehyde species and the ease of their conversion into free formaldehyde with increasing temperature. Consequently, it may be expected that the addition of a less polar solvent to an alcohol having a higher polarity (as well as the addition of water to methanol or ethanol) will be equivalent to an increase in temperature with respect to an increase in the concentration of the free form of dissolved formaldehyde.

Figure 3 illustrates the use of Equations (5) and (6) for describing the experimental dependences of the formation rates of 1,2-butanediol (curve *1*) in reactions 3 and 4 and propanal (curve *2*) in reaction 3a on the concentration of free formaldehyde in the 1-propanol–formaldehyde reacting system at a total formaldehyde concentration of 2.0 to 9.5 mol dm^{-3} and temperature of 413 K [8, 9, 41]. The concentration dependence of the propanal formation rate was described using the estimates of kinetic parameters obtained for the same dependence of the 1,2-butanediol formation rate.

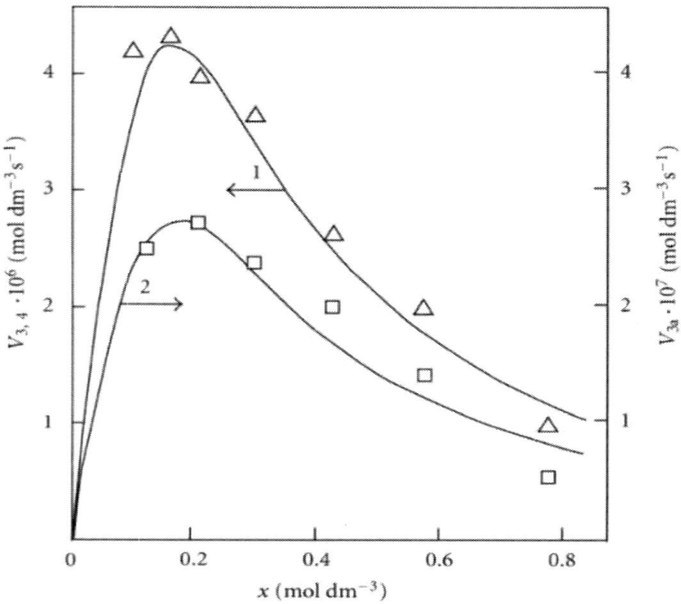

Figure 3. Reconstruction of the functional dependence (curves) of the product formation rates $V_{3,4}$ and V_{3a} on the concentration x of free formaldehyde (model optimization with respect to the parameters α, β and k_2) from empirical data (symbols) for the 1-propanol–formaldehyde system at 413 K [8, 9, 41]: (*1*, △) calculation using Equation (5), standard deviation of $S_Y = 2.20 \times 10^{-7}$; (*2*, □) calculation using Equation (6), $S_Y = 2.38 \times 10^{-8}$.

We considered these data more reliable for the reason that the carbonyl compounds forming in the alcohol–formaldehyde systems can react with the alcohol and this reaction depends considerably on the temperature and acidity of the medium [27]. The mathematical modeling of the process was carried out using a ^{137}Cs γ-radiation dose rate of P = 0.8 Gy s^{-1} [32, 41], a total initiation yield of $G(CH_3CH_2\dot{C}HOH)$ = 9.0 particles per 100 eV [8, 9] (V_1 = 4.07×10^{-7} mol dm^{-3} s^{-1}), and $2k_5 = 4.7 \times 10^9$ dm^3 mol^{-1} s^{-1}). The following values of the parameters were obtained: $\alpha = 0.36 \pm 0.07$, $\beta = 0.25 \pm 0.05$ mol dm^{-3}, and $k_2 = (6.0 \pm 1.4) \times 10^3$ dm^3 mol^{-1} s^{-1}.

Note that, as compared to the yields of 1,2-propanediol in the γ-radiolysis of the ethanol–formaldehyde system, the yields of 2,3-butanediol in the γ-radiolysis of the ethanol–acetaldehyde system are one order of magnitude lower [41]. Using data from [8, 9], it can be demonstrated that, at 433 K, the double bond of 2-propen-1-ol accepts the 1-hydroxyethyl radical 3.4 times more efficiently than the double bond of formaldehyde [42].

Addition of Hydroxymethyl Radicals

The addition of hydroxymethyl radicals to the carbon atom at the double bond of free formaldehyde molecules in methanol, initiated by the free-radical mechanism, results in the chain formation of ethanediol [34]. In this case, reaction 3a in *Scheme 2* is the reverse of reaction 2, the 1-hydroxyalkyl radical $^{\bullet}R_{(-H)}OH$ is the hydroxymethyl radical $^{\bullet}CH_2OH$, so reaction 3b is eliminated ($k_{3b} = 0$), and reaction 5 yields an additional amount of ethanediol *via* the dimerization of chain-carrier hydroxymethyl radicals (their disproportionation can practically be ignored [43]). The scheme of these reactions is presented in [35]. The experimental dependence of the ethanediol radiation-chemical yields on formaldehyde concentration in γ-radiolysis of methanol–formaldehyde system at 373–473 K is shown in Figure 4.

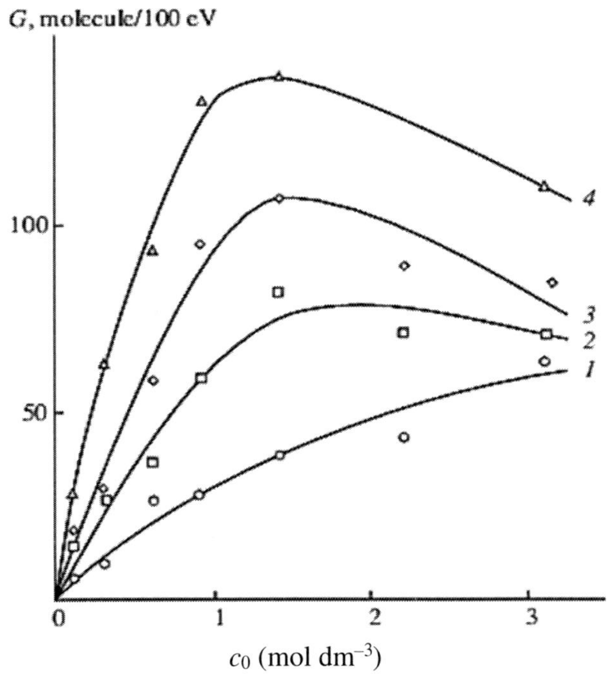

Figure 4. Dependence of the ethanediol yield G on the total concentration c_0 of formaldehyde upon γ-radiolysis of the methanol–formaldehyde system at temperatures of (*1*) 373, (*2*) 423, (*3*) 448, and (*4*) 473 K [34].

The maximum yield of the product reached 139 molecules per 100 eV (144 mol 10^{-7} J) at a formaldehyde concentration of 1.4 mol dm^{-3} in methanol and a temperature of 473 K. The dependence of the yield on the concentration at 373 K does not have a maximum even at the highest (3.1 mol dm^{-3}) concentration of formaldehyde, whereas maxima appear on the curves in the formaldehyde concentration region about 1.4 mol dm^{-3} at higher temperatures. The maximum in these curves arises from the competition between methanol and free formaldehyde ($H_2C{=}O$) for reacting with the adduct radical $HOCH_2CH_2O^{\bullet}$. The effective activation energy of ethylene glycol formation is 25 ± 3 kJ mol^{-1} [35].

The rate equation for ethanediol formation by the chain mechanism in reaction 3 and by the nonchain mechanism in reactions 4 and 5 in the methanol–formaldehyde system has a complicated form[3] as compared to Equation (1) for the formation rate of the other 1,2-alkanediols [34]:

$$V_{3,4,5}(CH_2OH)_2 = V_1[f(\alpha l + x)k_2x + V_1 2k_5(\alpha l + \beta + x)^2]/f^2 , \qquad (9)$$

where $f = k_2x^2 + (\alpha l + \beta + x)\sqrt{2k_5V_1}$.

If the rate of ethanediol formation by the dimerization mechanism in reaction 5 is ignored for the reason that it is small as compared to the total rate of ethanediol formation in reactions 3 and 4, Equation (9) will be identical to Equation (5). After the numerator and denominator on the right-hand side of Equation (5) are divided by $k_{-2} \equiv k_{3a}$, one can replace k_2 in this equation with $K_2 = k_2/k_{-2}$, which is the equilibrium constant for the reverse of reaction 2. Ignoring the reverse of reaction 2 ($k_{3a} = 0$, $\beta = 0$) makes Equation (5) identical to Equation (4) for *Scheme 1* at $k_{3b} = 0$ (see Section 2.1). In this case, the rate constant k_2 is effective.

Addition to Oxygen

The addition of a free radical or an atom to one of the two multiply bonded atoms of the oxygen molecule yields a peroxyl free radical and thus initiates oxidation, which is the basic process of chemical evolution. The peroxyl free radical then abstracts the most labile atom from a molecule of the compound being oxidized or decomposes to turn into a molecule of an oxidation product.

[3] In an earlier publication [8], this equation does not take into account reaction 3a.

The only reaction that can compete with these two reactions at the chain evolution stage is the addition of the peroxyl radical to the oxygen molecule (provided that the oxygen concentration is sufficiently high). This reaction yields a secondary, tetraoxyalkyl, 1:2 adduct radical, which is the heaviest and the largest among the reactants. It is less reactive than the primary, 1:1 peroxyl adduct radical and, as a consequence, does not participate in further chain propagation. At moderate temperatures, the reaction proceeds *via* a nonbranched-chain mechanism.

Addition of Hydrocarbon Free Radicals

Usually, the convex curve of the hydrocarbon (RH) autooxidation rate as a function of the partial pressure of oxygen ascends up to some limit and then flattens out [6]. When this is the case, the oxidation kinetics is satisfactorily describable in terms of the conventional reaction scheme [2, 5, 6, 16, 44, 45], which involves two types of free radicals. These are the hydrocarbon radical $R^{\bullet}$ (addend radical) and the addition product $RO_2^{\bullet}$ (1:1 adduct radical). However, the existing mechanisms are inapplicable to the cases in which the rate of initiated oxidation as a function of the oxygen concentration has a maximum (Figures 4, 5) [46, 47]. Such dependences can be described in terms of the competition kinetics of free-radical chain addition, whose reaction scheme involves not only the above two types of free radicals, but also the $RO_4^{\bullet}$ radical (1:2 adduct) inhibiting the chain process [13, 14].

Scheme 3

Chain initiation:

1. $I \xrightarrow{2k_1} 2R_0^{\bullet};$

1a. $R_0^{\bullet} + RH \xrightarrow{k_{1a}} R_0H + R^{\bullet}.$

Chain propagation:

2. $R^{\bullet} + O_2 \xrightarrow{k_2} RO_2^{\bullet};$

3. $RO_2^{\bullet} + RH \xrightarrow{k_3} RO_2H + R^{\bullet}$ (or $ROH + RO^{\bullet}$);

3a. $RO_2^{\bullet} \xrightarrow{k_{3a}} R'_{(-H)}HO + R''O^{\bullet}$

(or $R_{(-2H)}HO + {}^{\bullet}OH$);

3b. $R''O^{\bullet}(\text{or } RO^{\bullet}) + RH \xrightarrow{k_{3b}} R''OH\ (\text{or } ROH) + R^{\bullet}$

(or ${}^{\bullet}OH + RH \xrightarrow{k_{3b}} H_2O + R^{\bullet}$).

Inhibition:

4. $RO_2^{\bullet} + O_2 \xrightarrow{k_4} RO_4^{\bullet}$.

Chain termination:

5. $2R^{\bullet} \xrightarrow{2k_5} RR$

(or $R_{(-2H)}H + RH$);

6. $R^{\bullet} + RO_4^{\bullet} \xrightarrow{k_6} RH + R_{(-2H)}HO + O_3$

(or: $ROH + R_{(-2H)}HO + O_2$,

$ROR + O_3$,

$RO_2R + O_2$);

7. $2\,RO_4^{\bullet} \xrightarrow{2k_7} RO_2R + 2O_3$.

The only difference between the kinetic model of oxidation represented by *Scheme 3* and the kinetic model of the chain addition of 1-hydroxyalkyl radicals to the free (unsolvated) form of formaldehyde in non-methanolic alcohol–formaldehyde systems [8, 9] (*Scheme 2*, Section 3.1) is that in the former does not include the formation of the molecular 1:1 adduct *via* reaction 4.

The decomposition of the initiator I in reaction 1 yields a reactive $R_0^{\bullet}$ radical, which turns into the ultimate product R_0H *via* reaction 1a, generating an alkyl radical $R^{\bullet}$, which participates in chain propagation. In reaction 2, the addition of the free radical $R^{\bullet}$ to the oxygen molecule yields a reactive alkylperoxyl 1:1 adduct radical $RO_2^{\bullet}$ [45], which possesses increased energy owing to the energy released upon the conversion of the O=O bond into the ordinary bond RO–O$^{\bullet}$ (for addition in the gas phase under standard conditions, this energy is 115–130 kJ mol^{-1} for C_1–C_4 alkyl radicals [1, 2, 4] and 73 kJ mol^{-1} for the allyl radical [4]).

Because of this, the adduct radical can decompose (reaction 3a) or react with some neighbor molecule (reaction 3 or 4) on the spot, without diffusing in the solution and, accordingly, without entering into any chain termination reaction.

In reaction 3, the interaction between the radical adduct $RO_2^{\bullet}$ and the hydrocarbon molecule RH yields, *via* a chain mechanism, the alkyl hydroperoxide RO_2H (this reaction regenerates the chain carrier $R^{\bullet}$ and, under certain conditions, can be viewed as being reversible [2]) or the alcohol ROH (this is followed by the regeneration of $R^{\bullet}$ *via* reaction 3b). The latter (alternative) pathway of reaction 3 consists of four steps, namely, the breaking of old bonds and the formation of two new bonds in the reacting structures. In reaction 3a, the isomerization and decomposition of the alkylperoxyl radical adduct $RO_2^{\bullet}$ with O–O and C–O or C–H bond breaking take place [6, 44], yielding the carbonyl compound $R'_{(-H)}HO$ or $R_{(-2H)}HO$. Reaction 3b produces the alcohol R″OH or water and regenerates the free radical $R^{\bullet}$ (here, R′ and R″ are radicals having a smaller number of carbon atoms than R). As follows from the above scheme of the process, consecutive reactions 3a and 3b (whose rates are equal within the quasi-steady-state treatment), in which the highly reactive fragment, oxyl radical $R''O^{\bullet}$ (or $^{\bullet}OH$) forms and then disappears, respectively, can be represented as a single, combined bimolecular reaction 3a,b occurring in a "cage" of solvent molecules. Likewise, the alternative (parenthesized) pathways of reactions 3 and 3b, which involve the alkoxyl radical $RO^{\bullet}$, can formally be treated as having equal rates.

For simple alkyl C_1–C_4 radicals R, the pathway of reaction 3 leading to the alkyl hydroperoxide RO_2H is endothermic (ΔH°_{298} = 30–80 kJ mol^{-1}) and the alternative pathway yielding the alcohol ROH is exothermic (ΔH°_{298} = –120 to –190 kJ mol^{-1}), while the parallel reaction 3a, which yields a carbonyl compound and the alkoxyl radical $R''O^{\bullet}$ or the hydroxyl radical $^{\bullet}OH$, is exothermic in both cases (ΔH°_{298} = –80 to –130 kJ mol^{-1}), as also is reaction 3b (ΔH°_{298} = –10 to –120 kJ mol^{-1}), consecutive to reaction 3a, according to thermochemical data for the gas phase [2–4]. In reaction 4, which is competing with (parallel to) reactions 3 and 3a (chain propagation through the reactive radical $R^{\bullet}$), the resulting low-reactive radical that does not participate in further chain propagation and inhibits the chain process is supposed to be the alkyltetraoxyl 1:2 radical adduct[4,5] $RO_4^{\bullet}$, which has the largest weight and size. This radical is possibly stabilized by a weak intramolecular H···O hydrogen bond [54] shaping it into a six-membered cyclic structure[6] (seven-membered cyclic structure in the case of aromatic and certain branched acyclic hydrocarbons) [56, 57]:

4. 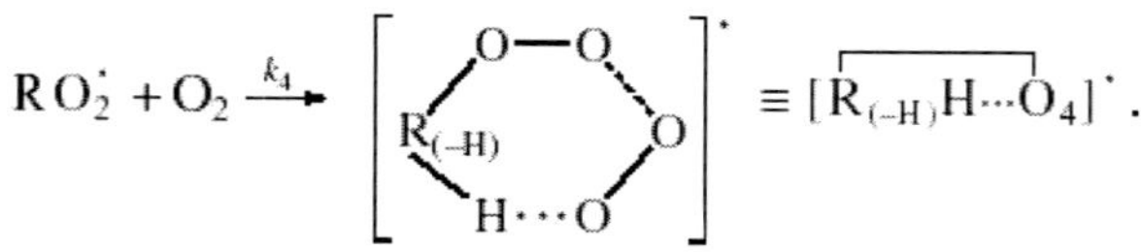

Reaction 4 in the case of the methylperoxyl radical $CH_3O_2^{\bullet}$ adding to the oxygen molecule to yield the methyltetraoxyl radical $CH_3O_4^{\bullet}$ takes place in the gas phase, with heat absorption equal to 110.0 ± 18.6 kJ mol^{-1} [49] (without the energy of the possible formation of a hydrogen bond taken into account). The exothermic reactions 6 and 7, in which the radical $R^{\bullet}$ or $RO_4^{\bullet}$ undergoes disproportionation, include the isomerization and decomposition of the $RO_4^{\bullet}$ radical.[7] The latter process is likely accompanied by chemiluminescence typical of hydrocarbon oxidation [52].

---

[4] It is hypothesized that raising the oxygen concentration in the *o*-xylene–oxygen system can lead to the formation of the $[RO^{\bullet}\cdots O_2]$ intermediate complex [46] similar to the $[ROO^{\bullet}\cdots(\pi\text{-bond})RH]$ complex between the alkylperoxyl 1:1 adduct radical and an unsaturated hydrocarbon suggested in this work. The electronic structure of the π-complexes is considered elsewhere [48].

[5] Thermochemical data are available for some polyoxyl free radicals (the enthalpy of formation of the methyltetraoxyl radical without the energy of the possible intramolecular hydrogen bond H···O taken into account is $\Delta H_f{}^{\circ}{}_{298}(CH_3O_4\cdot) = 121.3 \pm 15.3$ kJ mol^{-1}) and polyoxides ($\Delta H_f{}^{\circ}{}_{298}(CH_3O_4H) = -21.0 \pm 9$ kJ mol^{-1}) [49]. These data were obtained using the group contribution approach. Some physicochemical and geometric parameters were calculated for the methyl hydrotetraoxide molecule as a model compound [50–52]. The IR spectra of dimethyl tetraoxide with isotopically labeled groups in Ar–O_2 matrices were also reported [53]. For reliable determination of the number of oxygen atoms in an oxygen-containing species, it is necessary to use IR and EPR spectroscopy in combination with the isotope tracer method [53].

[6] The $\overline{R_{(-H)}H\cdots O(R)O}_3$ ring consisting of the same six atoms (C, H, and 4O), presumably with a hydrogen bond [6], also forms in the transition state of the dimerization of primary and secondary alkylperoxyl radicals $RO_2^{\bullet}$ *via* the Russell mechanism [5, 55].

[7] Taking into account the principle of detailed balance for the various pathways of formation of products, whose numbers in the elementary reaction should not exceed three for possible involvement in the triple collisions in the case of the reverse reaction, since the probability of simultaneous interaction of four particles is negligible. This principle was not taken into account in the author's works on oxidation until 2012.

These reactions regenerate oxygen as O_2 molecules (including singlet oxygen[8] [52, 59]) and, partially, as O_3 molecules and yield the carbonyl compound $R_{(-2H)}HO$ (possibly in the triplet excited state [52]). Depending on the decomposition pathway, the other possible products are the alcohol ROH, the ether ROR, and the alkyl peroxide RO_2R. It is likely that the isomerization and decomposition of the $RO_4^{\bullet}$ radical *via* reactions 6 and 7 can take place through the breaking of a C–C bond to yield carbonyl compounds, alcohols, ethers, and organic peroxides containing fewer carbon atoms than the initial hydrocarbon, as in the case of the alkylperoxyl radical $RO_2^{\bullet}$ in reaction 3a. At later stages of oxidation and sufficiently high temperatures, the resulting aldehydes can be further oxidized into respective carboxylic acids. They can also react with molecular oxygen so that a C–H bond in the aldehyde molecule breaks to yield two free radicals ($HO_2^{\bullet}$ and $^{\bullet}R'_{(-H)}O$ or $^{\bullet}R_{(-2H)}O$). This process, like possible ozone decomposition yielding an $^{\bullet}O^{\bullet}$ atom or peroxide decomposition with O–O bond breaking, leads to degenerate chain branching [6].

The equations describing the formation rates of molecular products at the chain propagation and termination stages of the above reaction scheme, set up using the quasi-steady-state treatment, appear as follows:

$$V_3(RO_2H;\, ROH) = V_1 \alpha l k_2 x / f \tag{10}$$

$$= V_1 \alpha l x / f_m \,, \tag{10a}$$

$$V_{3a}(R'_{(-H)}HO;\, R_{(-2H)}HO) = V_{3b}(R''OH;\, H_2O)$$

$$= V_1 \beta\, k_2 x / f \tag{11}$$

$$= V_1 \beta\, x / f_m \,, \tag{11a}$$

$$V_5 = V_1^2\, 2k_5 \left(\alpha l + \beta + x\right)^2 / f^2 \,, \tag{12}$$

$$2V_6 = 2V_1 \sqrt{2k_5 V_1} \left(\alpha l + \beta + x\right) k_2 x^2 / f^2 \,, \tag{13}$$

$$V_7 = V_1 \left(k_2 x^2\right)^2 / f^2 \,, \tag{14}$$

[8] Note that the alkylperoxyl radicals $RO_2^{\bullet}$ are effective quenchers of singlet oxygen $O_2(a^1\Delta_g)$ [58].

where V_1 is the initiation rate, l = [RH] and x = [O_2] are the molar concentrations of the starting components ($l >> x$), $\alpha = k_3/k_4$ and $\beta = k_{3a}/k_4$ (mol dm^{-3}) are the ratios of the rate constants of the competing (parallel) reactions,

$$k_2 = (\alpha l_m + \beta)\sqrt{2k_5V_1}/x_m^2$$

is the rate constant of the addition of the alkyl radical $R^{\bullet}$ to the oxygen molecule (reaction 2) as determined by solving the quadratic equation following from the rate function extremum condition $\partial V_{3,3a}/\partial x = 0$, l_m and x_m are the values of l and x at the maximum point of the function, $f = k_2x^2 + (\alpha l + \beta + x)\sqrt{2k_5V_1}$ and $f_m = x^2 + (\alpha l + \beta + x)x_m^2/(\alpha l_m + \beta)$.

The ratios of the rates of the competing reactions are $V_3/V_4 = \alpha l/x$ and $V_{3a}/V_4 = \beta/x$, and the chain length is $\nu = (V_3 + V_{3a})/V_1$. Equation (11) is identical to Equation (6). Equations (10a) and (11a) were obtained by replacing the rate constant k_2 in Equations (10) and (11) with its analytical expression (for reducing the number of unknown parameters to be determined directly).

For $\alpha l >> \beta$ ($V_3 >> V_{3a}$), when the total yield of alkyl hydroperoxides and alcohols having the same number of carbon atoms as the initial compound far exceeds the yield of carbonyl compounds, as in the case of the oxidation of some hydrocarbons, the parameter β in Equations (10) and (10a) can be neglected ($\beta = 0$) and these equations become identical to Equations (3) and (3a) with the corresponding analytical expression for k_2.

In the alternative kinetic model of oxidation, whose chain termination stage involves, in place of $R^{\bullet}$ (*Scheme 3*), $RO_2^{\bullet}$ radicals reacting with one another and with $RO_4^{\bullet}$ radicals, the dependences of the chain formation rates of the products on the oxygen concentration x derived by the same method have no maximum:

$$V_3 = V_1k_3l/(k_4x + \sqrt{2k_5V_1}) \text{ and } V_{3a} = V_1k_{3a}/(k_4x + \sqrt{2k_5V_1}).$$

In the kinetic model of oxidation that does not include the competing reaction 4 ($k_4 = 0$) and involves the radicals $R^{\bullet}$ and $RO_2^{\bullet}$ (the latter instead of $RO_4^{\bullet}$ in *Scheme 3*) in reactions 5–7, the reaction rate functions V_3 and V_{3a} obtained in the same way, are fractional rational functions in the form of $a_0x/(b_0x + c_0)$, where a_0, b_0, and c_0 are coefficients having no extremum. For a similar kinetic model in which reactions 3a,b and 4 appearing in the above scheme are missing ($k_{3a} = k_4 = 0$), Walling [5], using the quasi-steady-state treatment in the long kinetic chain approximation, when it can be assumed that $V_2 = V_3$, without using the substitution $k_6 = \sqrt{2k_5 2k_7}$ [5, 6, 16] (as distinct from this work), found that $V_2 = V_3$ is an irrational function of x: $a_1x/\sqrt{b_1x^2 + c_1x + d_1}$ where a_1, b_1, c_1, and d_1 are coefficients. Again, this function has no maximum with respect to the concentration of any of the two components.

Thus, of the three kinetic models of oxidation mathematically analyzed above –which involve the radicals $R^{\bullet}$ and $RO_2^{\bullet}$ in three types of quadratic-law chain termination reactions (reactions 5–7) and are variants of the conventional model [2, 5, 6, 16, 44, 45] – the last two lead to an oxidation rate versus oxygen concentration curve that emanates from the origin of coordinates, is convex upward, and has an asymptote parallel to the abscissa axis. Such monotonic dependences are observed when the oxygen solubility in the liquid is limited under given experimental conditions and the oxygen concentration attained[9] is $[O_2]_{top} \leq x_m$.

Unlike the conventional model, the above kinetic model of free-radical nonbranched-chain oxidation, which includes the pairs of competing reactions 3–4 and 3a–4 (*Scheme 3*), allows us to describe the nonmonotonic (peaking) dependence of the oxidation rate on the oxygen concentration (Figure 5). In this oxidation model, as the oxygen concentration in the binary system is increased, oxygen begins to act as an oxidation autoinhibitor or an antioxidant *via* the further oxidation of the alkylperoxyl 1:1 adduct radical $RO_2^{\bullet}$ into the low-reactive 1:2 adduct radical $RO_4^{\bullet}$ (reactions 4 and 6 lead to inefficient consumption of the free radicals $RO_2^{\bullet}$ and $R^{\bullet}$ and cause shortening of the kinetic chains). The optimum oxygen concentration x_m, at which the oxidation rate is the highest, can be calculated using kinetic Equations (10a) and (11a) and Equation (3a) with $\beta = 0$ or the corresponding analytical expression for k_2. In the familiar monograph "Chain Reactions" by Semenov [60], it is noted that raising the oxygen concentration when it is already sufficient usually slows down the oxidation process by shortening the chains. The existence of the upper (second) ignition limit in oxidation is due to chain termination in the bulk through triple collisions between an active species of the chain reaction and two oxygen molecules (at sufficiently high oxygen partial pressures).

[9] The oxygen concentration attained in the liquid may be below the thermodynamically equilibrium oxygen concentration because of diffusion limitations hampering the establishment of the gas–liquid saturated solution equilibrium under given experimental conditions (for example, when the gas is bubbled through the liquid) or because the Henry law is violated for the given gas–liquid system under real conditions.

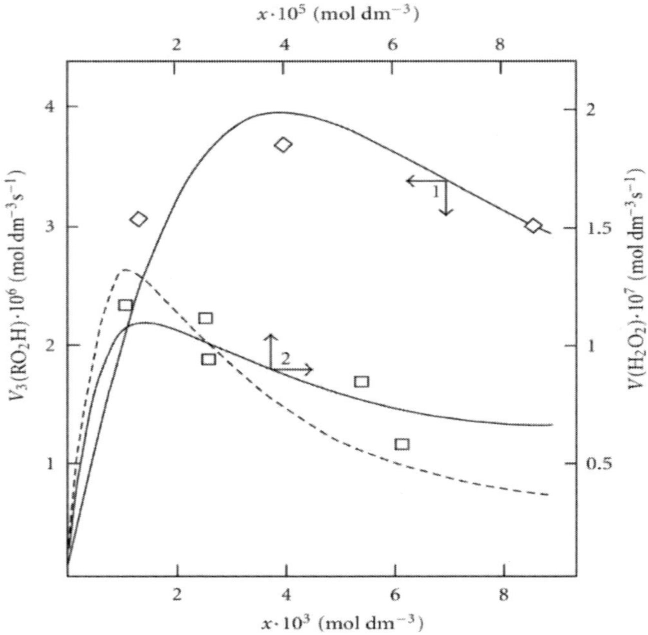

Figure 5. (*1*, ◇) Reconstruction of the functional dependence of the 2-methylbenzyl hydroperoxide formation rate $V_3(RO_2H)$ on the dissolved oxygen concentration x from empirical data (points) using Equation (3a) (model optimization with respect to the parameter α) for the *o*-xylene–oxygen system at 373 K [46] (standard deviation of $S_Y = 5.37 \times 10^{-7}$. (*2*, □) Reconstruction of the functional dependence of the total hydrogen peroxide formation rate $V_{3,7}(H_2O_2)$ on the dissolved oxygen concentration x from empirical data (symbols) using Equations (3a) and (14a) with $\beta = 0$ (model optimization with respect to the parameter α) for the γ-radiolysis of water saturated with hydrogen and containing different amounts of oxygen at 296 K [63] ($S_Y = 1.13 \times 10^{-8}$). The dashed curve described $V_3(H_2O_2)$ as a function of the oxygen concentration x based on Equation (3a) (model optimization with respect to α) and the experimental data of curve *2* ($S_Y = 1.73 \times 10^{-8}$).

In the gas phase at atmospheric pressure, the number of triple collisions is roughly estimated to be 10^3 times smaller than the number of binary collisions (and the probability of a reaction taking place depends on the specificity of the action of the third particle) [60]. Note that in the case of gas-phase oxidation of hydrogen at low pressures of 25-77 Pa and a temperature of 77 K [47] when triple collisions are unlikely, the dependence of the rate of hydrogen peroxide formation on oxygen concentration (the rate of passing of molecular oxygen *via* the reaction tube) also has a pronounced maximum (see curves *3* and *4* in Figure 6) that indicates a chemical mechanism providing the appearance of a maximum (see reaction 4 of *Scheme 4*).

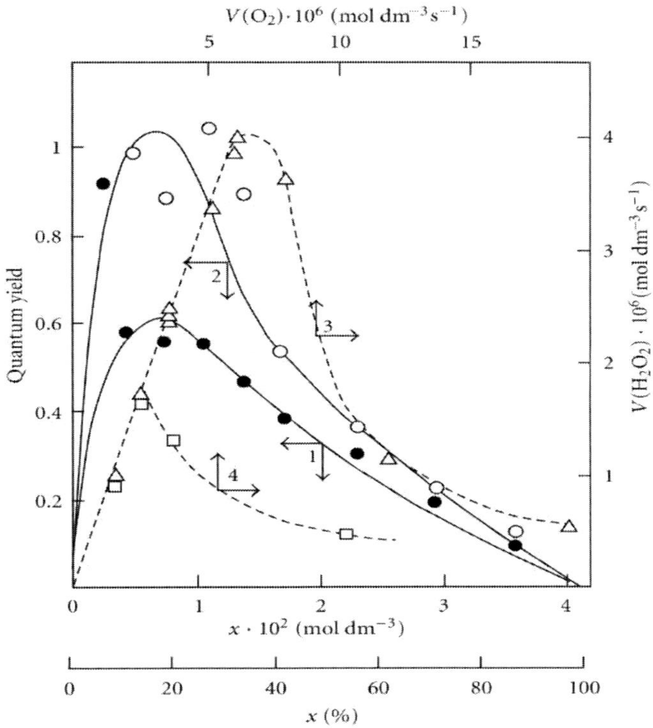

Figure 6. (*1*, *2*) Quantum yields of (*1*, ●) hydrogen peroxide and (*2*, ○) water resulting from the photochemical oxidation of hydrogen in the hydrogen–oxygen system as a function of the oxygen concentration *x* (light wavelength of 171.9–172.5 nm, total pressure of 10^5 Pa, room temperature [64]). (*3*, *4*) Hydrogen peroxide formation rate $V(H_2O_2)$ (dashed curves) as a function of the rate $V(O_2)$ at which molecular oxygen is passed through a gas-discharge tube filled with (*3*, △) atomic and (*4*, □) molecular hydrogen. Atomic hydrogen was obtained from molecular hydrogen in the gas-discharge tube before the measurements (total pressure of 25–77 Pa, temperature of 77 K [47]). The symbols represent experimental data.

Curve *1* in Figure 5 illustrates the fit between Equation (3a) at $\alpha l >> \beta$ and experimental data for the radiation-induced oxidation of *o*-xylene in the liquid phase at 373 K in the case of 2-methylbenzyl hydroperoxide forming much more rapidly than *o*-tolualdehyde ($V_3 >> V_{3a}$ and $\alpha l >> \beta$) [46]. The oxygen concentration limit in *o*-xylene is reached at an oxygen concentration of $[O_2]_{top} > x_m$, which corresponds to the third experimental point [46]. The oxygen concentration was calculated from the oxygen solubility in liquid xylene at 373 K [61]. The following quantities were used in this mathematical description: ^{60}Co γ-radiation dose rate of $P = 2.18$ Gy s^{-1} and total initiation yield of $G(o\text{-}CH_3C_6H_4\dot{C}H_2) = 2.6$ particles per 100 eV of the energy absorbed by the solution [46]; $V_1 = 4.73 \times 10^{-7}$ mol dm^{-3} s^{-1}, and $2k_5 = 1.15 \times 10^{10}$ dm^3 mol^{-1} s^{-1}. The resulting value of the parameter α is $(9.0 \pm 1.8) \times 10^{-3}$; hence, $k_2 = (3.2 \pm 0.8) \times 10^5$ dm^3 mol^{-1} s^{-1}. From the data presented in [62], it was estimated that $k_4 = k_3/\alpha = (5.2 \pm 1.2) \times 10^2$ dm^3 mol^{-1} s^{-1}.

Addition of the Hydrogen Atom

Several experimental findings concerning the autoinhibiting effect of an increasing oxygen concentration at modest temperatures on hydrogen oxidation both in the liquid phase [63] (Figure 5, curve *2*) and in the gas phase [47, 64, 65] (Figure 6), considered in our earlier work [13, 56, 57, 66], can also be explained in terms of the competition kinetics of free radical addition [14, 67]. Figure 6 shows that the quantum yields of hydrogen peroxide and water (of products of photochemical oxidation of hydrogen at atmospheric pressure and room temperature) are maximum in the region of small concentrations of oxygen in the hydrogen–oxygen system (curves *1* and *2*, respectively) [64].

Scheme 4

Nonbranched-chain oxidation of hydrogen and changes in enthalpy (ΔH°_{298}, kJ mol^{-1}) for elementary reactions.[10]

[10] According to Francisco and Williams [49], the enthalpy of formation ($\Delta H_f{}^{\circ}{}_{298}$) in the gas phase of $H^{\bullet}$, $HO^{\bullet}$, $HO_2^{\bullet}$, $HO_4^{\bullet}$ (the latter without the possible intramolecular hydrogen bond taken into account), O_3, H_2O [2], H_2O_2, and H_2O_4 is 218.0 ± 0.0, 39.0 ± 1.2, 12.6 ± 1.7, 122.6 ± 13.7, 143.1 ± 1.7, –241.8 ± 0.0, –136.0 ± 0, and –26.0 ± 9 kJ mol^{-1}, respectively. Calculations for the $HO_4^{\bullet}$ radical with a helical structure were carried out using the G2(MP2) method [68]. The stabilization energies of $HO_2^{\bullet}$, $HO_4^{\bullet}$, and $HO_3^{\bullet}$ were calculated in the same work to be 64.5 ± 0.1, 69.5 ± 0.8, and 88.5 ± 0.8 kJ mol^{-1}, respectively. The types of the O_4 molecular dimers, their IR spectra, and higher oxygen oligomers were reported [69, 70]. The structure and IR spectrum of the hypothetical cyclotetraoxygen molecule O_4, a species with a high-energy density, were calculated by the CCSD method, and its enthalpy of formation was estimated [71]. The photochemical properties of O_4 and the van der Waals nature of the O_2–O_2 bond were investigated [72, 73]. The most stable geometry of the dimer is two O_2 molecules parallel to one another. The O_4 molecule was identified by NR mass spectrometry [74].

Chain initiation:

$$1.\ H_2 \xrightarrow{h\nu,\gamma} 2H^{\bullet}, \qquad \Delta H^{o}_{298} = 436.0 \pm 0.0.$$

Chain propagation:

$$2.\ H^{\bullet} + O_2 \xrightarrow{k_2} HO_2^{\bullet}, \qquad \Delta H^{o}_{298} = -205.4 \pm 1.7;$$

$$3.\ HO_2^{\bullet} + H_2 \xrightarrow{k_3} H_2O + HO^{\bullet}, \qquad \Delta H^{o}_{298} = -215.4 \pm 2.9$$

$$(\text{or } H_2O_2 + H^{\bullet}), \qquad \Delta H^{o}_{298} = 69.4 \pm 1.7;$$

$$3'.\ HO^{\bullet} + H_2 \xrightarrow{k_{3'}} H_2O + H^{\bullet}, \qquad \Delta H^{o}_{298} = -62.8 \pm 1.2.$$

Inhibition:

$$4.\ HO_2^{\bullet} + O_2 \xrightarrow{k_4} HO_4^{\bullet}, \qquad \Delta H^{o}_{298} = 110.0 \pm 15.4.$$

Chain termination:

$$5.\ 2H^{\bullet}\ (+\,M) \xrightarrow{2k_5} H_2\ (+\,M), \qquad \Delta H^{o}_{298} = -436.0 \pm 0.0;$$

$$6.\ H^{\bullet} + HO_4^{\bullet} \xrightarrow{k_6} H_2O_2 + O_2, \qquad \Delta H^{o}_{298} = -476.6 \pm 13.7$$

$$(\text{or: } H_2O + O_3, \qquad \Delta H^{o}_{298} = -439.3 \pm 15.4,$$

$$H_2 + 2O_2), \qquad \Delta H^{o}_{298} = -340.6 \pm 13.7;$$

$$7.\ 2\,HO_4^{\bullet} \xrightarrow{2k_7} H_2O_2 + 2O_3, \qquad \Delta H^{o}_{298} = -95.0 \pm 30.8.$$

The hydroperoxyl free radical $HO_2^{\bullet}$ [75–78] resulting from reaction 2 possesses an increased energy due to the energy released in the conversion of the O=O multiple bond into the HO–O$^{\bullet}$ ordinary bond. Therefore, before its possible decomposition, it can interact with a hydrogen or oxygen molecule as the third body *via* parallel (competing) reactions 3 and 4, respectively. The hydroxyl radical HO$^{\bullet}$ that appears and disappears in consecutive parallel reactions 3 (first variant) and 3′ possesses additional energy owing to the exothermicity of the first variant of reaction 3, whose heat is distributed between the two products. As a consequence, this radical has a sufficiently high reactivity not to accumulate in the system during these reactions, whose rates are equal ($V_3 = V_{3'}$) under quasi-steady-state conditions, according to the above scheme. Parallel reactions 3 (second, parenthesized variant) and 3′ regenerate hydrogen atoms. It is assumed [56, 57] that the hydrotetraoxyl radical $HO_4^{\bullet}$ (first reported in [79, 80]) resulting from endothermic reaction 4, which is responsible for the peak in the experimental rate curve (Figure 5, curve *2*), is closed into a five-membered [OO—H···OO]˙ cycle due to weak intramolecular hydrogen bonding [54, 81]. This structure imparts additional stability to this radical and makes it least reactive.

The $HO_4^{\bullet}$ radical was discovered by Staehelin *et al.* [82] in a pulsed radiolysis study of ozone degradation in water; its UV spectrum with an absorption maximum at 260 nm ($\varepsilon(HO_4^{\bullet})_{280\,nm}$= 320 ±15 m^2 mol^{-1}) was reported. The spectrum of the $HO_4^{\bullet}$ radical is similar to that of ozone, but the molar absorption coefficient $\varepsilon(HO_4^{\bullet})_{\lambda_{max}}$ of the former is almost two times larger [82]. The assumption about the cyclic structure of the $HO_4^{\bullet}$ radical can stem from the fact that its mean lifetime in water at 294 K, which is (3.6 ± 0.4) × 10^{-5} s (as estimated [66] from the value of 1/*k* for the reaction: $HO_4^{\bullet} \xrightarrow{k} HO_2^{\bullet} + O_2$ [82]), is 3.9 times longer than that of the linear $HO_3^{\bullet}$ radical [68, 83] estimated in the same way [66] for the same conditions [84], (9.1 ± 0.9) × 10^{-6} s.

MP2/6-311++G** calculations using the Gaussian-98 program confirmed that the cyclic structure of $HO_4^{\bullet}$ [85] is energetically more favorable than the helical structure [68] (the difference in energy is 4.8–7.3 kJ mol^{-1}, depending on the computational method and the basis set).[11] For example, with the MP2(full)/6-31G(d) method, the difference between the full energies of the cyclic and acyclic $HO_4^{\bullet}$ conformers with their zero-point energies (ZPE) values taken into account (which reduces the energy difference by 1.1 kJ mol^{-1}) is –5.1 kJ mol^{-1} and the entropy of the acyclic-to-cyclic $HO_4^{\bullet}$ transition is ΔS°_{298}= –1.6 kJ mol^{-1} K^{-1}. Therefore, under standard conditions, $HO_4^{\bullet}$ can

[11] There were calculations for the two conformers (*cis* and *trans)* of the $HO_4^{\bullet}$ radical [86] using large scale *ab initio* methods and density functional techniques with extended basis sets. Both conformers have a nearly planar geometry with respect to the four oxygen atoms and present an unusually long central O–O bond. The most stable conformer of $HO_4^{\bullet}$ radical is the *cis* one, which is computed to be endothermic with respect to $HO_2^{\bullet}(X^2A'') + O_2(X^3\Sigma_g^-)$ at 0 K.

exist in both forms, but the cyclic structure is obviously dominant (87%, K_{eq} = 6.5) [85].

Reaction 4 and, to a much lesser degree, reaction 6 inhibit the chain process, because they lead to inefficient consumption of its main participants, $HO_2^{\bullet}$ and $H^{\bullet}$.

The hydrogen molecule that results from reaction 5 in the gas bulk possesses an excess energy, and, to acquire stability within the approximation used in this work, it should have time for deactivation *via* collision with a particle M capable of accepting the excess energy [87]. To simplify the form of the kinetic equations, it was assumed that the rate of the bimolecular deactivation of the molecule substantially exceeds the rate of its monomolecular decomposition, which is the reverse of reaction 5 [2].

Reactions 6 and 7 (taking into account the principle of detailed balance for the various pathways) regenerate hydrogen and oxygen (in the form of $O_2(X^3\Sigma_g^-)$ molecules, including the singlet states with $\Delta H^{o}_{f298}(O_2,\ a^1\Delta_g)$ = 94.3 kJ mol^{-1} [49, 70] and, $\Delta H^{o}_{f298}(O_2,\ b^1\Sigma_g^+)$ = 161.4 kJ mol^{-1} [70], which are deactivated by collisions, and in the form of O_3) and yield hydrogen peroxide or water *via* a nonchain mechanism, presumably through the intermediate formation of the unstable hydrogen tetraoxide molecule H_2O_4 [88].[12] Ozone does not interact with molecular hydrogen. At moderate temperatures, it decomposes fairly slowly, particularly in the presence of $O_2(X^3\Sigma_g^-)$ [70]. The reaction of ozone with $H^{\bullet}$ atoms, which is not impossible, results in their replacement with $HO^{\bullet}$ radicals. The relative contributions from reactions 6 and 7 to the process kinetics can be roughly estimated from the corresponding enthalpy increments (*Scheme 4*).

When there is no excess hydrogen in the hydrogen–oxygen system and the homomolecular dimer O_4 [71–74, 89, 90], which exists at low concentrations (depending on the pressure and temperature) in equilibrium with O_2 [70], can directly capture the $H^{\bullet}$ atom to yield the heteronuclear cluster $HO_4^{\bullet}$,[13] which is more stable than O_4 [70] and cannot abstract a hydrogen atom

---

[12] The planar, six-atom, cyclic, hydrogen-bonded dimer $(HO_2^{\bullet})_2$ was calculated using quantum chemical methods (B3LYP density functional theory) [88]. The hydrogen bond energy is 47.7 and 49.4 kJ mol^{-1} at 298 K for the triplet and singlet states of the dimer, respectively.

[13] It is impossible to make a sharp distinction between the two-step bimolecular interaction of three species *via* the equilibrium formation of the labile intermediate O_4 and the elementary trimolecular reaction $O_2 + O_2 + H^{\bullet} \rightarrow HO_4^{\bullet}$.

from the hydrogen molecule, nonchain hydrogen oxidation will occur to give molecular oxidation products *via* the disproportionation of free radicals.

The low-reactive hydrotetraoxyl radical $HO_4^{\bullet}$ [82], which presumably has a high-energy density [71], may be an intermediate in the efficient absorption and conversion of biologically hazardous UV radiation energy in the Earth's upper atmosphere. The potential energy surface for the atmospheric reaction $HO^{\bullet} + O_3$, in which the adduct $HO_4^{\bullet}$ (^2A) was considered as an intermediate, was calculated by the double many-body expansion (DMBE) method [91]. From this standpoint, the following reactions are possible in the upper troposphere, as well as in the lower and middle stratosphere, where most of the ozone layer is situated (altitude of 16–30 km, temperature of 217–227 K, pressure of 1.0×10^4–1.2×10^3 Pa [92]; the corresponding ΔH°_{298} reaction values are given in kJ mol^{-1} [49]):

8. $H_2O(vapor) + h\nu \rightarrow H^{\bullet} + HO^{\bullet}$ [92];

9. $HO^{\bullet} + O_3 \rightarrow HO_4^{\bullet}$ [80, 82, 91], $\Delta H^{o}_{298} = -59.5$;

10. $HO_4^{\bullet} \rightarrow HO_2^{\bullet} + O_2(X^3\Sigma_g^-)$ [82, 91], $\Delta H^{o}_{298} = -110.0$

(or $HO_2^{\bullet} + O_2(a^1\Delta_g)$), $\Delta H^{o}_{298} = -15.7$.

The $HO_4^{\bullet}$ radical can disappear *via* disproportionation with a molecule, free radical, or atom in addition to dissociation. Note that emission from $O_2(a^1\Delta_g)$ and $O_2(b^1\Sigma_g^+)$ is observed at altitudes of 30–80 and 40–130 km, respectively [93].

Staehelin *et al.* [82] pointed out that, in natural systems in which the concentrations of intermediates are often very low, kinetic chains in chain reactions can be very long in the absence of scavengers since the rates of the chain termination reactions decrease with decreasing concentrations of the intermediates according to a quadratic law, whereas the rates of the chain propagation reactions decrease according to a linear law.

The kinetic description of the noncatalytic oxidation of hydrogen, including in an inert medium [87], in terms of the simplified scheme of free-radical nonbranched-chain reactions (*Scheme 4*), which considers only quadratic-law chain termination and ignores the surface effects [47], at moderate temperatures and pressures, in the absence of transitions to unsteady-state critical regimes, and at a substantial excess of the hydrogen concentration over the oxygen concentration was obtained by means of quasi-steady-state treatment, as in the previous studies on the kinetics of the branched-chain free-radical oxidation of hydrogen [76], even though the applicability of this method in the latter case under unsteady states conditions was insufficiently substantiated. The method was used with the following condition:[14] $k_6 = \sqrt{2k_5 2k_7}$ (see Introduction). The equation for the rate of the chain formation of hydrogen peroxide and water, $V_3(H_2O_2; H_2O) = V_{3'}(H_2O)$, *via* reactions 3 and 3′ is identical to Equation (3, 3a) with the corresponding analytical expression for k_2. The ratio of the rates of the competing reactions is $V_3/V_4 = \alpha l/x$, and the chain length is $\nu = V_3/V_1$. The rates of nonchain formation of hydrogen peroxide and water *via* reactions (6) and (7) – quadratic-law chain termination – are identical to Equations. (13) and (14) provided that $\beta = 0$. In these equations, l and x are the molar concentrations of hydrogen and oxygen ($l >> x$), l_m and x_m are the respective concentrations at the maximum point of the function, V_1 is the rate of initiation (reaction 1), $\alpha = k_3/k_4$, the rate constant:

$$k_2 = \alpha l_m \sqrt{2k_5 V_1} / x_m^2$$

is derived from the condition $\partial V_3/\partial x = 0$, and $2k_5$ is the rate constant of reaction 5 (hydrogen atom recombination), which is considered as bimolecular within the given approximation.[15]

In the case of nonchain hydrogen oxidation *via* the above addition reaction ($H^{\bullet} + O_4 \xrightarrow{k_{add}} HO_4^{\bullet}$), the formation rates of the molecular oxidation products in reactions 6 and 7 (*Scheme 4*, $k_2 = k_3 = k_4 = 0$) are defined by modified Equations (13) and (14) in which $\beta = 0$, $(\alpha l + x)$ is replaced with 1, and k_2 is replaced with $k_{add}K_{eq}$ ($k_{add}K_{eq}$ is the effective rate constant of $H^{\bullet}$ addition to the O_4 dimer, $K_{eq} = k/k'$ is the equilibrium constant of the reversible reaction: $2O_2 \underset{k'}{\overset{k}{\Leftrightarrow}} O_4$ with $k' >> k_{add}[H^{\bullet}]$). The formation rates of the stable

[14] For example, the ratio of the rate constants of the bimolecular disproportionation and dimerization of free radicals at room temperature is $k(HO^{\bullet} + HO_2^{\bullet})/[2k(2HO^{\bullet})2k(2HO_2^{\bullet})]^{0.5} = 2.8$ in the atmosphere [92] and $k(H^{\bullet} + HO^{\bullet})/[2k(2H^{\bullet})2k(2HO^{\bullet})]^{0.5} = 1.5$ in water [94]. These values that are fairly close to unity.

[15] This rate constant in the case of the pulsed radiolysis of ammonia–oxygen (+ argon) gaseous mixtures at a total pressure of 10^5 Pa and a temperature of 349 K was calculated to be 1.6×10^8 dm^3 mol^{-1} s^{-1} [65] (a similar value of this constant for the gas phase was reported in an earlier publication [95]). Pagsberg *et al.* [65] found that the dependence of the yield of the intermediate $HO^{\bullet}$ on the oxygen concentration has a maximum close to 5×10^{-4} mol dm^{-3}. In the computer simulation of the process, they considered the strongly exothermic reaction $HO_2^{\bullet} + NH_3 \rightarrow H_2O + {}^{\bullet}NHOH$, which is similar to reaction 3 in *Scheme 4*, whereas the competing reaction 4 was not taken into account.

products of nonchain oxidation ($k_3 = 0$), provided that either reactions (2) and (4) or reaction (2) alone ($k_4 = 0$) occurs (*Scheme 4*; in the latter case, reactions 6 and 7 involve the $HO_2^{\bullet}$ radical rather than $HO_4^{\bullet}$), are given by modified Equations (13) and (14) with $\beta = 0$, $(\alpha l + x)$ replaced with 1, and x^2 replaced with x.

Note that, if in *Scheme 4* chain initiation *via* reaction 1 is due to the interaction between molecular hydrogen and molecular oxygen yielding the hydroxyl radical $HO^{\bullet}$ instead of $H^{\bullet}$ atoms and if this radical reacts with an oxygen molecule (reaction 4) to form the hydrotrioxyl radical $HO_3^{\bullet}$ (which was obtained in the gas phase by neutralization reionization (NR) mass spectrometry [83] and has a lifetime of $>10^{-6}$ s at 298 K) and chain termination takes place *via* reactions 5–7 involving the $HO^{\bullet}$ and $HO_3^{\bullet}$ radicals instead of $H^{\bullet}$ and $HO_4^{\bullet}$, respectively, the expressions for the water chain formation rates derived in the same way will appear as a rational function of the oxygen concentration x without a maximum: $V_{3'}(H_2O) = V_1 k_{3'} l / (k_4 x + \sqrt{2k_5 V_1})$.

Curve 2 in Figure 5 describes, in terms of the overall equation: $V_{3,7} = V_1 x(\alpha l f_m + x^3)/f_m^2$ for the rates of reactions 3 and 7 (which was derived from Equations 3a and 14, respectively, the latter in the form of $V_7 = V_1 x^4 / f_m^2$ (14a) [96] in which k_2 is replaced with its analytical expression derived from Equation (10) with $\beta = 0$ everywhere), the dependence of the hydrogen peroxide formation rate (minus the rate $V_{H_2O_2} = 5.19 \times 10^{-8}$ mol dm^{-3} s^{-1} of the primary formation of hydrogen peroxide after completion of the reactions in spurs) on the concentration of dissolved oxygen during the γ-radiolysis of water saturated with hydrogen (at the initial concentration 7×10^{-4} mol dm^{-3}) at 296 K [63]. These data were calculated in the present work from the initial slopes of hydrogen peroxide buildup versus dose curves for a ^{60}Co γ-radiation dose rate of $P = 0.67$ Gy s^{-1} and absorbed doses of $D \cong 22.5$–304.0 Gy.

The following values of the primary radiation-chemical yield G (species per 100 eV of energy absorbed) for water γ-radiolysis products in the bulk of solution at pH 4–9 and room temperature were used (taking into account that $V = GP$ and $V_1 = G_H P$): $G_{H_2O_2} = 0.75$ and $G_H = 0.6$ (initiation yield; see below) [94]; $V_1 = 4.15 \times 10^{-8}$ mol dm^{-3} s^{-1}; $2k_5 = 2.0 \times 10^{10}$ dm^3 mol^{-1} s^{-1} [94]. As can be seen from Figure 5, the best description of the data with an increase in the oxygen concentration in water is attained when the rate V_7 of the formation of hydrogen peroxide *via* the nonchain mechanism in the chain termination reaction 7 (curve *1*, $\alpha = (8.5 \pm 2) \times 10^{-2}$) is taken into account in addition to the rate V_3 of the chain formation of this product *via* the propagation reaction

3 (dashed curve *2*, $\alpha = 0.11 \pm 0.026$). The rate constant of addition reaction 2 determined from α is substantially underestimated: $k_2 = 1.34 \times 10^7$ (*vs.* 2.0×10^{10} [94]) $dm^3\ mol^{-1}\ s^{-1}$. The difference can be due to the fact that the radiation-chemical specifics of the process were not considered in the kinetic description of the experimental data. These include oxygen consumption *via* reactions that are not involved in the hydrogen oxidation scheme [66, 97, 98] and reverse reactions resulting in the decomposition of hydrogen peroxide by intermediate products of water radiolysis (e^-_{aq}, $H^\bullet$, $HO^\bullet$), with the major role played by the hydrated electron [94].

General Scheme of the Addition of Free Radicals to Molecules of Alkenes, Formaldehyde, and Oxygen

The general scheme of the nonbranched-chain addition of a free radical from a saturated compound to an alkene (and its functionalized derivative), formaldehyde, or dioxygen (which can add an unsaturated radical as well) in liquid homogeneous binary systems of these components includes the following reactions [57, 97, 98].

Scheme 5

Initiation:

1. $I \xrightarrow{2k_1} 2R_0^\bullet$;

1a. $R_0^\bullet + R_1A \xrightarrow{k_{1a}} R_0A + R_1^\bullet$;

For addition to an alkene at comparable component concentrations,

1b. $R_0^\bullet + R_2B \xrightarrow{k_{1b}} R_0B + R_2^\bullet$.

Chain propagation:

2. $R_1^\bullet + R_2B \xrightarrow{k_2} R_3^\bullet$;

3. $R_3^{\bullet} + R_1A \xrightarrow{k_3} R_3A + R_1^{\bullet}$;

For addition to O_2 and the 1-hydroxyalkyl radical to CH_2O,

3a. $R_3^{\bullet} \xrightarrow{k_{3a}} R'R''CO + R_4^{\bullet}$;

3b. $R_4^{\bullet} + R_1A \xrightarrow{k_{3b}} R_4A + R_1^{\bullet}$.

Inhibition:

For addition to an alkene or CH_2O,

4. $R_3^{\bullet} + R_2B \xrightarrow{k_4} R_3B + R_2^{\bullet}$;

For addition to O_2,

4a. $R_3^{\bullet} + R_2B \xrightarrow{k_{4a}} R_{2a}^{\bullet}$.

Chain termination:

5. $2R_1^{\bullet} \xrightarrow{2k_5} \text{Prod}$;

6. $R_1^{\bullet} + R_{2(2a)}^{\bullet} \xrightarrow{k_6} \text{Prod}$;

7. $2R_{2(2a)}^{\bullet} \xrightarrow{2k_7} \text{Prod}$.

In this scheme, I is the initiator, for example, a peroxide [5, 17, 18, 29, 30]; $R_0^{\bullet}$ is any reactive radical (initiator); A is an atom of hydrogen [2, 5, 6, 17, 18, 22–24, 29–32] or halogen [2, 5, 19–21]; B is an atom of hydrogen [5, 17–21, 23, 24, 29–32], halogen [22], or oxygen (in oxidation) [2, 5, 6, 16, 44–46]; $R_1^{\bullet}$ is a radical such as $^{\bullet}PCl_2$ [19], $^{\bullet}CCl_3$ [20], an alkyl [2, 5, 6, 21], a 1-hydroxyalkyl [5, 6, 17, 22–24, 29, 32], or a similar functionalized radical [5] (addend); $R_2^{\bullet}$ is the formyl [8, 9, 29], an alkenyl (propenyl or higher) [2, 5, 17–22], a 1-hydroxyalkenyl [5, 17, 18, 23, 24], or a similar functionalized low-reactive radical [5, 18] (inhibitor) or the oxygen atom (in oxidation) [2, 5, 6, 13, 14, 16, 44–46, 56, 57, 96–98]; $R_{2a}^{\bullet}$ is the low-reactive alkyltetraoxyl 1:2

adduct radical $RO_4^{\bullet}$ [13, 14, 56, 57, 96–98] (inhibitor); $R_3^{\bullet}$ is the active 1:1 adduct radical; $R_4^{\bullet}$ is an active fragment radical, such as hydroxymethyl [8, 9, 12, 29, 32], an alkoxyl radical, or hydroxyl (in oxidation) [2, 5, 6, 13, 14, 16, 44, 46, 56, 57, 96–98]; R_0A, R_0B, R_1A, and R_4A are saturated molecules; R_2B is an unsaturated molecule, *viz*., an alkene [2, 5, 11, 17–22], formaldehyde [8, 9, 12, 29–32], or dioxygen (in oxidation) [2, 5, 6, 13, 14, 16, 44–46, 56, 57, 96–98]; R′R″CO is a carbonyl compound, *viz*., aldehyde [2, 6, 8, 9, 12, 14, 29–32, 44] or ketone [2, 6, 14, 29, 32, 44]; R_3A and R_3B are molecular products (1:1 adducts); and Prod stands for molecular products of the dimerization and disproportionation of free radicals.

The chain evolution stage of *Scheme 5* includes consecutive reactions 2, 3; 2, 3a; and 3a, 3b; parallel (competitive) reactions 3, 3a; 3, 3b; 3, 4 (or 4a); and 3a, 4 (or 4a); and consecutive-parallel reactions 2 and 4 (or 4a). Addition to alkenes is described by reactions 1–3, 4, and 5–7 and the corresponding rate Equations (1)–(4a). Addition to the carbonyl carbon atom of the free (unsolvated) form of formaldehyde is represented by reactions 1, 1a, 2–4 (the main products are a 1,2-alkanediol, a carbonyl compound, and methanol), and 5–7 and is described by Equations (5) and (6). In the case of hydroxymethyl addition, the process includes reactions 1, 1a, 2, 3, 5a, 4 (the main product is ethanediol), and 5–7 and is described by Equation (9). If the nonchain formation of ethanediol in reaction 5 is ignored, the process is described by Equation (5). Addition to the oxygen molecule is described by reactions 1, 1a, 2–3b, 4a (the main products are an alkyl hydroperoxide, alcohols, carbonyl compounds, and water), and 5–7 and Equations (10) and (11).

The main molecular products of the chain process – R_3A, R′R″CO, and R_4A – result from reactions 3, 3a, and 3b – chain propagation through the reactive free radical $R_1^{\bullet}$ or $R_4^{\bullet}$, R′R″CO. The competing reaction 4, which opposes this chain propagation, yields the by-product R_3B a nonchain mechanism. The rate of formation of the products is a complicated function of the formation rates ($V_{3a} = V_{3b}$) and disappearance rates of the free radicals $R_1^{\bullet}$ and $R_{2(2a)}^{\bullet}$: $V(R_3A, R'R''CO, R_4A, R_3B) = V_2 = V_3 + V_{3a} + V_{4(4a)} = (V_{1a} + V_3 + V_{3b} - V_5) - (V_{1b} + V_{4(4a)} - V_7)$. The rates of reactions 5–7 at $k_{1b} = 0$ ($[R_1A] >> [R_2B]$) are given by Equations (12)–(14). The rate ratios of the competing reactions are $V_3/V_{4(4a)} = \alpha l/x$ and $V_{3a}/V_{4(4a)} = \beta/x$ (where $\alpha = k_3/k_{4(4a)}$, $\beta = k_{3a}/k_{4(4a)}$ mol dm^{-3}, and l and x are the molar concentrations of the reactants R_1A and R_2B, respectively), and the chain length is $\nu = (V_3 + V_{3a})/V_1$. Unlike the dependences of the rates of reactions 4a (or 4 at $k_{1b} = 0$, with $V_{4(4a)} \leq V_1$), 5, and 7 (for the last two – Equations (12) and (14)), the dependences of the rates V of reactions 3, 3a,b, 4 (at $k_{1b} \neq 0$), and 6 (Equations (1), (3)–(6), (10), (11), and (13)) on x have a maximum. Reaction 1b, which competes with reaction 1a, gives rise to a maximum in the dependence described by Equation (2), whereas reaction 4 or 4a, competing with reactions 3 and 3a,b, is responsible for the maxima in the dependences defined by Equations (1), (3)–(6) or (10) and (11). The low-reactive radicals $R_2^{\bullet}$[16] and $R_{2a}^{\bullet}$, resulting from reactions 4 and 4a, inhibit the nonbranched-chain addition of $R_1^{\bullet}$ to alkenes (or formaldehyde) and dioxygen, respectively. Reaction 4a leads to non-productive loss of $R_3^{\bullet}$ adduct radicals.

For approximate estimation of the parameters of the kinetic Equations (3), (4), (10), and (11), Equation (4) under the conditions (a) $k_2x^2 \ll (\alpha l + x)\sqrt{2k_5V_1}$ (ascending branch of a peaked curve) and (b) $k_2x^2 \gg (\alpha l + x)\sqrt{2k_5V_1}$ (descending branch) is transformed into simple functions (direct and inverse proportionality, respectively) of the concentration x of the unsaturated compound, which allow tentative estimates of the parameters k_2 and α to be derived from the experimental product formation rate V provided that V_1 and $2k_5$ are known:

$$V_{3,4} = \sqrt{V_1}k_2x \big/ \varphi\sqrt{2k_5}\,, \tag{15}$$

$$V_{3,4} = (V_1/\varphi)[(\alpha l/x) + 1], \tag{16}$$

where $\varphi = 1$ under conditions (a) and (b) and $\varphi = 2$ at the point of maximum (where $k_2x^2 \cong (\alpha l + x)\sqrt{2k_5V_1}$). Equations (10) and (11) under the condition $k_2x^2 \gg (\alpha l + \beta + x)\sqrt{2k_5V_1}$ (descending branch of a peaked curve) can be transformed into Equations (17) and (18), respectively, which express the simple, inversely proportional dependences of reaction rates on x and provide tentative estimates of α and β:

$$V_3 = V_1\alpha l/\varphi x, \tag{17}$$

$$V_{3a} = V_1\beta/\varphi x, \tag{18}$$

[16] The stabilization energy of the low-reactive free radicals $CH_2{=}C(CH_3)\dot{C}H_2$, $CH_2{=}CH\dot{C}HOH$, and $H\dot{C}{=}O$ in the standard state in the gas phase is –52.0, –42.1, and –24.3 kJ mol^{-1}, respectively [4, 99].

where $\varphi = 2$ at the point of maximum (where $k_2x^2 \cong (al + \beta + x)\sqrt{2k_5V_1}$) and $\varphi = 1$ for the descending branch of the curve. Equation (3) for $V_{3,4}$ under condition (b) transforms into Equation (17).

For radiation-chemical processes, the rates V in the kinetic equations should be replaced with radiation-chemical yields G using the necessary unit conversion factors and the relationships $V = GP$ and $V_1 = \varepsilon_1 G(\mathrm{R}_1^\bullet)P$, where P is the dose rate, ε_1 is the electron fraction of the saturated component R_1A in the reaction system [100], and $G(\mathrm{R}_1^\bullet)$ is the initial yield of the chain-carrier free radicals (addends) – initiation yield [39, 94].

Conclusion

In summary, the material on the kinetics of nonbranched-chain addition of free saturated radicals to multiple bonds of alkene (or its derivative), formaldehyde, or oxygen molecules makes it possible to describe, using rate Equations (1)–(6), (9)–(11) obtained by quasi-steady-state treatment, experimental dependences with a maximum of the formation rates of molecular 1:1 adducts on the concentration of an unsaturated compound over the entire region of its change in binary reaction systems consisting of saturated and unsaturated components (Figures 1, 3, 5).

The proposed addition mechanism involves the reaction of a free 1:1 adduct radical with an unsaturated molecule yielding a low-reactive free radical (the reaction 4 competing with the chain propagation reactions in *Schemes 1–5*). In such reaction systems, the unsaturated compound is both a reactant and an autoinhibitor, specifically, a source of low-reactive free radicals shortening kinetic chains. The progressive inhibition of the nonbranched-chain processes, which takes place as the concentration of the unsaturated compound is raised (after the maximum process rate is reached), can be an element of the self-regulation of the natural processes that returns them to the stable steady state.

A similar description is applicable to the nonbranched-chain free-radical hydrogen oxidation in water at 296 K [63] (Figure 5, curve *2*). Using the hydrogen oxidation mechanism considered here, it has been demonstrated that, in the Earth's upper atmosphere, the decomposition of O_3 in its reaction with the $HO^\bullet$ radical can occur *via* the addition of the latter to the ozone molecule, yielding the $HO_4^\bullet$ radical, which is capable of efficiently absorbing UV radiation [82].

The optimum concentration x_m of the unsaturated component in the binary system at which the process rate is maximal can be derived with the help of obtained kinetic Equations (3a), (4a), (10a), and (11a) or from the corresponding analytical expressions for k_2 if other parameters are known. This opens a way to intensification of some technological processes that are based on the addition of free radicals to the double bonds of unsaturated molecules and occur *via* a nonbranched-chain mechanism through the formation of 1:1 adducts.

References

[1] Gurvich, L. V., Karachevtsev, G. V., Kondrat'ev, V. N., Lebedev, Yu. A., Medvedev, V. A., Potapov, V. K., Khodeev, Yu. S. *Energii razryva khimicheskikh svyazei. Potentsialy ionizatsii i srodstvo k elektronu* [*Bond Dissociation Energies, Ionization Potentials, and Electron Affinity*]; Kondrat'ev, V. N., Ed; Nauka: Moscow, 1974, 351 pp.

[2] Benson, S. W. *Thermochemical Kinetics: Methods for the Estimation of Thermochemical Data and Rate Parameters*; 2nd Edition; Wiley: New York, 1976, 320 pp.

[3] Pedley, J. B., Naylor, R. D., Kirby, S. P. *Thermochemical Data of Organic Compounds*; 2nd Edition; Chapman & Hall: London, 1986, 792 pp.

[4] Orlov, Yu. D., Lebedev, Yu. A., Saifullin, I. Sh. *Termokhimiya organicheskikh svobodnykh radikalov* [*Thermochemistry of Organic Free Radicals*]; Kutepov, A. M., Ed., Nauka: Moscow, 2001, 304 pp.

[5] Walling, Ch. *Free Radicals in Solution*; Wiley: New York; Chapman & Hall: London, 1957, 631 pp.

[6] Emanuel, N. M., Denisov, E. T., Maizus, Z. K. *Tsepnye reaktsii okisleniya uglevodorodov v zhidkoi faze* [*Chain Oxidation Reactions of Hydrocarbons in the Liquid Phase*]; Nauka: Moscow, 1965, 375 pp.

[7] Poluektov, V. A., Babkina, E. I., Begishev, I. R. On the Dependence of the Rate of a Chain Reaction on the Reactant Ratio. *Dokady Akademii Nauk SSSR*. 1974, *215*(3), 649–652.

[8] Silaev, M. M., Bugaenko, L. T. Mathematical Simulation of the Kinetics of Radiation Induced Hydroxyalkylation of Aliphatic Saturated Alcohols. *Radiation Physics and Chemistry*. 1992, *40*(1), 1–10.

[9] Silaev, M. M., Bugaenko, L. T. Kinetics of the Addition of α-Hydroxyalkyl Radicals to 2-Propen-1-ol and Formaldehyde. *Kinetics and Katalysis*. 1994, *35*(4), 509–513.

[10] Silaev, M. M. Competition Kinetics of Nonbranched Chain Processes of Free-Radical Addition to Double Bonds of Molecules with the Formation of 1:1 Adducts. *Kinetica i Kataliz*. 1999, *40*(2), 281–284 [English Translation in: *Kinetics and Catalysis*. 1999, *40*(2), 256–259].

[11] Silaev, M. M. Simulation of the Nonbranched-Chain Addition of Saturated Free Radicals to Alkenes and Their Derivatives Yielding 1:1 Adducts. *Teoreticheskie Osnovy Khimicheskoi Tekhnologii.* 2007, *41*(3), 280–295 [English Translation in: *Theoretical Foundations of Chemical Engineering.* 2007, *41*(3), 273–278].

[12] Silaev, M. M. Simulation of Nonbranched Chain Processes for Producing 1,2-Alkanediols in Alcohol–Formaldehyde Systems. *Teoreticheskie Osnovy Khimicheskoi Tekhnologii.* 2007, *41*(4), 379–384 [English Translation in: *Thoretical Foundations Chemical Engineering.* 2007, *41*(4), 357–361].

[13] Silaev, M. M. A New Competitive Kinetic Model of Radical Chain Oxidation: Oxygen as an Autoinhibitor. *Biofizika.* 2001, *46*(2), 203–209 [English Translation in: *Biophysics.* 2001, *46*(2), 202–207].

[14] Silaev, M. M. Simulation of the Initiated Addition of Hydrocarbon Free Radicals and Hydrogen Atoms to Oxygen *via* a Nonbranched Chain Mechanism. *Teoreticheskie Osnovy Khimicheskoi Tekhnologii.* 2007, *41*(6), 634–642 [English Translation in: *Theoretical Foundation of Chemical Engineering.* 2007, *41*(6), 831–838].

[15] Bard, Y. *Nonlinear Parameter Estimation*; Academic Press: New York, 1974, 341 pp.

[16] Bateman, L. Alkene Oxidation. *Quarterly Reviews.* 1954, *8*(2), 1954, 147–167.

[17] Urry, W. H., Stacey, F. W., Huyser, E. S., Juveland, O. O. The Peroxide- and Light-Induced Additions of Alcohols to Alkenes. *Journal of the American Chemical Society.* 1954, *76*(2), 450–455.

[18] Urry, W.H., Juveland, O.O. Free Radical Additions of Amines to Alkenes. *Journal of the American Chemical Society.* 1958, *80*(13), 3322–3328.

[19] Shostenko, A. G., Zagorets, P. A., Dodonov, A. M., Greish, A. A. γ-Radiation-Induced Addition of Phosphorus Trichloride to Isobutylene. *Khimiya Vysokikh Energii.* 1970, *4*(4), 357.

[20] Kim, V., Shostenko, A. G., Gasparyan, M. D. Reactivity of Polychloroalkyl Radicals in the Telomerization of CCl_4 with 1-Propene and 2-Methyl-1-Propene (in Russian). *Reaction Kinetics and Catalysis Letters.* 1979, *12*(4), 479–484.

[21] Myshkin, V. E., Shostenko, A. G., Zagorets, P. A., Markova, K. G., Pchelkin, A. I. Determination of Absolute Rate Constants for the Addition of the Ethyl Radical to Alkenes. *Teoreticheskaya i Eksperimental'naya Khimiya.* 1977, *13*(2), 266–271.

[22] Zamyslov, R. A., Shostenko, A. G., Dobrov, I. V., Tarasova, N. P. Kinetics of γ-Radiation-Induced Reactions of 2-Propanol with Trifluoropropene and Hexafluoropropene. *Kinetika i Katalyz.* 1987, *28*(4), 977–979.

[23] Silaev, M. M. Dependence of Radiation-chemical γ-Diol Yields on the 2-Propen-1-ol Concentration in the Radiolysis of Aliphatic Saturated C_1-C_3 Alcohol + 2-Propen-1-ol Systems. *Khimiya Vysokikh Energii.* 1990, *24*(3), 282–283.

[24] Silaev, M. M. γ-Diol Formation *via* the Autooxidation of 2-Propen-1-ol Solutions in Saturated Alcohols. *Vestnik Moskovskogo Universiteta; Ser. 2: Khimiya.* 1994, *35*(1), 40–42.

[25] Bugaenko, L. T., Kuzmin, M. G., Polak, L. S. *High-Energy Chemistry;* Horwood Hall: New York, 1993, p. 112.

[26] Thomas, J. K. Pulse Radiolysis of Aqueous Solutions of Methyl Iodide and Methyl Bromide. The Reactions of Iodine Atoms and Methyl Radicals in Water. *The Journal of the Physical Chemistry.* 1967, *71*(6), 1919–1925.

[27] Walker, J. F. Formaldehyde; Reinhold: New York, 1953 [English Translation under the title Formal'degid; Goskhimizdat: Moscow, 1957, p. 106].

[28] Silaev, M. M., Rudnev, A. V., Kalyazin, E. P. Formaldehyde. III. Concentration of Free Formaldehyde as a Function of Temperature, Polarity of Solvents, and Total Concentration of Formaldehyde in Solution. *Zhurnal Fizicheskoi Khimii.* 1979, *53*(7), 1647–1651.

[29] Oyama, M. A Free-Radical Reaction of Primary and Secondary Alcohols with Formaldehyde. *The Journal of Organic Chemistry.* 1965, *30*(7), 2429–2432.

[30] Nikishin, G. I., Lefor, D., Vorob'ev, E. D. Free Radical Reaction of Primary Alcohols with Formaldehyde. *Izvestiya Akademii Nauk SSSR; Ser.: Khimiya.* 1966, (7), 1271–1272.

[31] Dzhurinskaya, M. B., Rudnev, A. V., Kalyazin, E. P. High Temperature UV Photolysis of Formaldehyde in Liquid Methanol. *Vestnik Moskovskogo Universiteta; Ser. 2: Khimiya.* 1984, *25*(2), 173–176.

[32] Kalyazin, E. P., Petryaev, E. P., Shadyro, O. I. Reaction between Oxyalkyl Radicals and Aldehydes. *Zhurnal Organicheskoi Khimii.* 1977, *13*(2), 293–295.

[33] Novoselov, A. I., Silaev, M. M., Bugaenko, L. T. Effect of Temperature on the Yields of Final Products in the γ-Radiolysis of Formaldehyde Solutions in C_1-C_3 Alkanols. *Khimiya Vysokich Energii.* 2004, *38*(4), 270–272 [English Translation in: *High Energy Chemistry.* 2004, *38*(4), 236–238].

[34] Novoselov, A. I., Silaev, M. M., Bugaenko, L. T. Dependence of Ethanediol Yield on Formaldehyde Concentration in γ-Radiolysis of Methanol–Formaldehyde System at 373–473 K. *Khimiya Vysokikh Energii.* 2008, *42*(1), 74–75 [English Translation in: *High Energy Chemistry.* 2008, *42*(1), 69–70].

[35] Novoselov, A. I., Silaev, M. M., Bugaenko, L. T. γ-Induced Single-Step Synthesis of Ethylene Glycol from Methanol–Formaldehyde Solution. *Theoreticheskie Osnovy Khimicheskoy Tekhnologii.* 2010, *44*(4), 450–453 [English Translation in: *Theoretical Foundation of Chemical Engineering.* 2010, *44*(4), 432–435].

[36] Novoselov, A. I., Silaev, M. M., Bugaenko, L. T. Dependence of 1,2-Propanediol Yield on Formaldehyde Concentration in γ-Radiolysis of Ethanol–Formaldehyde System at 373473 K. *Khimiya Vysokikh Energii.* 2007, *41*(1), 58 [English Translation in: *High Energy Chemistry.* 2007, *41*(1), 53].

[37] Pshezhetskii, S. Ya., Kotov, A. G., Milinchuk, V. K., Roginskii, V. A., Tupikov, V. I. *EPR svobodnykh radikalov v radiatsionnoi khimii* (ESR of Free Radicals in Radiation Chemistry; Khimiya: Moscow, 1972, 480 pp. [Translated into English by Shelnitz, P., Halstead Press: Wiley, New York, 1974, 446 pp]

[38] Silaev, M. M. Estimating the Solvent Concentration in Formaldehyde Solutions at Various Temperatures. *Zhurnal Fizicheskoy Khimii.* 1993, *67*(9), 1944.

[39] Silaev, M. M. Applied Aspects of the γ-Radiolysis of C_1-C_4 Alcohols and Binary Mixtures on Their Basis. *Khimiya Vysokikh Energii.* 2002, *36*(2), 97–101 [English Translation in: *High Energy Chemistry.* 2002, *36*(2), 70–74].

[40] Silaev, M. M., Bugaenko, L. T., Kalyazin, E. P. On the Possibility of Adequately Estimating the Rate Constants for the Reaction of Hydroxyalkyl Radicals with Each Other Using the Self-Diffusion Coefficients or Viscosities of the Corresponding Alcohols. *Vestnik Moskovskogo. Univiversiteta; Ser. 2: Khimiya*. 1986, *27*(4), 386–389.

[41] Shadyro, O. I. *Radiation-chemical Conversions of Aldehydes in Various Systems.* Ph.D. Thesis (Chemistry); Belarusian State University; Minsk, 1975, 24 pp.

[42] Silaev, M. M. Relative Reactivity of α-Hydroxyethyl Radicals for 2-Propene-1-ol and Formaldehyde Double-Bond Addition. *Vestnik Moskovskogo Universiteta; Ser. 2: Khimiya*. 1993, *34*(3), p. 311.

[43] Seki, H., Nagai, R., Imamura, M. γ-Radiolysis of a Binary Mixture of Methanol and Water. The Formation of Formaldehyde in the Radiolysis of Liquid Methanol. *Bulletin of the Chemical Society of Japan.* 1968, *41*(12), 2877–2881.

[44] Shtern, V. Ya. Mekhanizm okisleniya uglevodorodov v gazovoi faze [*Mechanism of the Gas-Phase Oxidation of Hydrocarbons*]; Akademiya Nauk SSSR: Moscow, 1960, 496 pp.

[45] Bäckström, H. L. J. Der Kettenmechanismus bei der Autoxydation von Aldehyden [*The chain mechanism in the autoxidation of aldehydes*]. *Zeitschrift für physikalische Chemie (B).* 1934, *25*(1–2), 99–121.

[46] Aliev, A. A., Saraeva, V. V. Isomerization of Peroxy Radicals Resulting from the Radiation-Induced Oxidation of *o*-Xylene. *Vestnik Moskovskogo Universiteta; Ser. 2: Khimiya.* 1983, *34*(4), 371–374.

[47] Badin, E. J. The Reaction between Atomic Hydrogen and Molecular Oxygen at Low Pressures. Surface Effects. *Journal of the American Chemistry Society.* 1948, *70*(11), 3651–3655.

[48] Buchachenko, A. L. *Kompleksy radikalov i molekulyarnogo kisloroda s organicheskimi molekulami* [*Complexes of Radicals and Dioxygen with Organic Molecules*]; Beletskaya, I. P., Ed., Nauka: Moscow, 1984, 158 pp.

[49] Francisco, J. S., Williams, I. H. The Thermochemistry of Polyoxides and Polyoxy Radicals. *International Journal of Chemical Kinetics.* 1988, *20*(6), 455–466.

[50] Kokorev, V. N., Vyshinskii, N. N., Maslennikov, V. P., Abronin, I. A., Zhidomirov, G. M., Aleksandrov, Yu. A. Electronic Structure and Chemical Reactions of Peroxides: I. MINDO/3 Calculation of the Geometry and Enthalpy of Formation of the Ground States of Organic and Organoelement Peroxides. *Zhurnal Strukturnoi Khimii*. 1981, *22*(4), 9–15.

[51] Dmitruk, A. F., Lobanov, V. V., Kholoimova, L. I. Role of Tetroxide Conformation in the Mechanism of Peroxy Radical Recombination. *Teoreticheskaya i Eksperimental'naya Khimiya.* 1986, *22*(3), 363–366.

[52] Belyakov, V. A., Vasil'ev, R. F., Ivanova, N. M., Minaev, B. F., Osyaeva, O. V., Fedorova, G. F. Electronic Model of the Excitation of Chemiluminescence in the Oxidation of Organic Compounds. *Izvestiya Akademii Nauk SSSR; Ser.: Fizika.* 1987, *51*(3), 540–547.

[53] Ase, P., Bock, W., Snelson, A. Alkylperoxy and Alkyl Radicals. 1. Infrared Spectra of CH_3O_2 and $CH_3O_4CH_3$ and the Ultraviolet Photolysis of CH_3O_2 in Argon + Oxygen Matrices. *The Journal of Physical Chemistry.* 1986, *90*(10), 2099–2109.

[54] Pimentel, G. C., McClellan, A. L. *The Hydrogen Bond*; Pauling, L., Ed., Freeman: San Francisco, 1960, p. 200.

[55] Russell, G. A. Deuterium-Isotope Effects in the Autooxidation of Aralkyl Hydrocarbons: Mechanism of the Interaction of Peroxy Radicals. *Journal of the American Chemical Society*. 1957, *79*(14), 3871–3877.

[56] Silaev, M. M. The Competition Kinetics of Nonbranched Chain Processes of Free-Radical Addition to Double Bonds of Molecules with the Formation of 1:1 Adducts and the Inhibition by the Substrate. *Oxidation Communication.* 1999, *22*(2), 159–170.

[57] Silaev, M. M. The Competition Kinetics of Radical-Chain Addition. *Zhurnal Fizicheskoi Khimii*. 1999, 73(7), 1180–1184 [English Translation in: *Russian Journal of Physical Chemistry*. 1999, *73*(7), 1050–1054].

[58] Darmanyan, A. P., Gregory, D. D., Guo, Y., Jenks, W. S., Burel, L., Eloy, D., Jardon, P. Quenching of Singlet Oxygen by Oxygen- and Sulfur-Centered Radicals: Evidence for Energy Transfer to Peroxy Radicals in Solution. *Journal of the American Chemistry Society*. 1998, *120*(2), 396–403.

[59] Kanofsky, J. R. Singlet Oxygen Production from the Reactions of Alkylperoxy Radicals. Evidence from 1268-nm Chemiluminescence. *The Journal of Organic Chemistry*. 1986, *51*(17), 3386–3388.

[60] Semenov, N. N. *Tsepnye reaktsii* [*Chain Reactions*]; Goskhimtekhizdat: Leningrad, 1934, pp. 241, 203.

[61] Reznikovskii, M., Tarasova, Z., Dogadkin, B. Oxygen Solubility in Some Organic Liquids. *Zhurnal Obshchei Khimii*. 1950, *20*(1), 63–67.

[62] Howard, J. A., Ingold, K. U. Absolute Rate Constants for Hydrocarbon Autooxidation. VI. Alkyl Aromatic and Alkeneic Hydrocarbons. *Canadian Journal of Chemistry*. 1967, *45*(8), 793–802.

[63] Barr, N. F., Allen, A. O. Hydrogen Atoms in the Radiolysis of Water. *The Journal of Physical Chemistry*. 1959, *63*(6), 928–931.

[64] Smith, H. A., Napravnik, A. Photochemical Oxidation of Hydrogen. *Journal of the American Chemistry Society*. 1940, *62*(1), 385–393.

[65] Pagsberg, P. B., Eriksen, J., Christensen, H. C. Pulse Radiolysis of Gaseous Ammonia–Oxygen Mixtures. *The Journal of Physical Chemistry*. 1979, *83*(5), 582–590.

[66] Silaev, M. M. Competitive Mechanism of the Nonbranched Radical Chain Oxidation of Hydrogen Involving the Free Cyclohydrotetraoxyl Radical $[\overline{OO\cdots H\cdots OO}]^{\bullet}$, Which Inhibits the Chain Process. *Khimiya Vysokikh Energii*. 37(1), 27–32 [English Translation in: *High Energy Chemistry*. 2003, *37*(1), 24–28].

[67] Silaev, M. M. Simulation of Initiated Nonbranched Chain Oxidation of Hydrogen: Oxygen as an Autoinhibitor. *Khimiya Vysokich Energii*. 2008, *42*(2), 124–129 [English Translation in: *High Energy Chemistry*. 2008, *42*(1), 95–100].

[68] McKay, D. J., Wright, J. S. How Long Can You Make an Oxygen Chain? *Journal of the American Chemistry Society*. 1998, *120*(5), 1003–1013.

[69] Lipikhin, N. P. Dimers, Clusters, and Cluster Ions of Oxygen in the Gas Phase. *Uspekhi Khimii*. 1975, *44*(8), 366–376.

[70] Razumovskii, S. D. *Kislorod – elementarnye formy i svoistva* [*Oxygen: Elementary Forms and Properties*]; Khimiya: Moscow, 1979, 302 pp.

[71] Dunn, K. M., Scuceria, G. E., Schaefer III, H. F. The infrared spectrum of cyclotetraoxygen, O_4: a theoretical investigation employing the single and double excitation coupled cluster method. *The Journal of Chemical Physics.* 1990, *92*(10), 6077–6080.

[72] Brown, L., Vaida, V. Photoreactivity of Oxygen Dimers in the Ultraviolet. *The Journal of Physical Chemistry*. 1996, *100*(19), 7849–7853.

[73] Aquilanti, V., Ascenzi, D., Bartolomei, M., Cappelletti, D., Cavalli, S., de Castro-Vitores, M., Pirani, F. Molecular Beam Scattering of Aligned Oxygen Molecules. The Nature of the Bond in the O_2–O_2 Dimer. *Journal of the American Chemistry Society*. 1999, *121*(46), 10794–1080.

[74] Cacace, F., de Petris, G., Troiani, A. Experimental Detection of Tetraoxygen. *Angewandte Chemie; Internation Edition* (in English). 2001, *40*(21), 4062–4065.

[75] Taylor, H. S. Photosensitisation and the Mechanism of Chemical Reactions (*Transactions of the Faraday Society*. 1926, *21*(63/3), 560–568).

[76] Nalbandyan, A. B., Voevodskii, V. V. *Mekhanizm okisleniya i goreniya vodoroda* [*Mechanism of Hydrogen Oxidation and Combustion*]; Kondrat'ev, V. N., Ed., Akademiya Nauk SSSR: Moscow–Leningrad, 1949, 180 pp.

[77] Foner, S. N., Hudson, R. L. Mass spectrometry of the HO_2 free radical. *The Journal of Chemical Physics*. 1962, *36*(10), 2681.

[78] Hochanadel, C. J., Ghormley, J. A., Ogren, P. J. Absorption Spectrum and Reaction Kinetics of the HO_2 Radical in the Gas Phase. *The Journal of Chemical Physics*. 1972, *56*(9), 4426–4432.

[79] Robertson, J. B. A Mass Spectral Search for H_2O_4 and HO_4 in a Gaseous Mixture Containing HO_2 and O_2. *Chemistry and Industry*. 1954, (48), 1485.

[80] Bahnemann, D., Hart, E. J. Rate Constants of the Reaction of the Hydrated Electron and Hydroxyl Radical with Ozone in Aqueous Solution. *The Journal of Physical Chemistry*. 1982, *86*(2), 252–255.

[81] *Vodorodnaya svyaz: Sbornik statei* [*The Hydrogen Bonding: Collection of Articles*]; Sokolov, N. D., Ed., Nauka: Moscow, 1981, 288 pp.

[82] Staehelin, J., Bühler, R. E., Hoigné, J. Ozone Decomposition in Water Studied by Pulse Radiolysis. 2. OH and HO_4 as Chain Intermediates. *The Journal of Physical Chemistry*. 1984, *88*(24), 5999–6004.

[83] Cacace, F., de Petris, G., Pepi, F., Troiani, A. Experimental Detection of Hydrogen Trioxide. *Science*. 1999, *285*(5424), 81–82.

[84] Bühler, R. F., Staehelin, J., Hoigné, J. Ozone Decomposition in Water Studied by Pulse Radiolysis. 1. HO_2/O_2^- and HO_3/O_3^- as Intermediates. *The Journal of Physical Chemistry*. 1984, *88*(12), 2560–2564.

[85] Trushkov, I. V., Silaev, M. M., Chuvylkin, N. D. Acyclic and Cyclic Forms of the Radicals $HO_4^\bullet$, $CH_3O_4^\bullet$, and $C_2H_5O_4^\bullet$: Ab Initio Quantum Chemical Calculations. *Izvestiya Akademii Nauk; Ser.: Khimiya*. 2009, (3), 479–482 [English Translation in: *Russian Chemical Bulletin; International Edition*. 2009, *58*(3), 489–492].

[86] Mansergas, A., Anglada, J. M., Olivella, S., Ruiz-López, M. F. On the Nature of the Unusually Long OO Bond in HO_3 and HO_4 Radicals. *Phys. Chem. Chem. Phys.* 2007, *9*(44), 5865–5873.

[87] Wong, W., Davis, D. D. A Flash Photolysis Resonance Fluorescence Study of the Reactions of Atomic Hydrogen and Molecular Oxygen: $H + O_2 + M \rightarrow HO_2 + M$. *International Journal of Chemical Kinetics*. 1974, *6*(3), 401–416.

[88] Xu, X., Muller, R. P., Goddard III, W. A. The Gas Phase Reaction of Singlet Dioxygen with Water: a Water-Catalyzed Mechanism. *Proceedings of the National Academy Sciences of the United States of America*. 2002, *99*(6), 3376–3381.

[89] Seidl, E. T., Schaefer III, H. F. Is There a Transition State for the Unimolecular Dissociation of Cyclotetraoxygen (O_4)? *The Journal of Chemical Physics*. 1992, *96*(2), 1176–1182.

[90] Hernández-Lamoneda, R., Ramírez-Solís, A. Reactivity and Electronic States of O_4 along Minimum Energy Paths. *The Journal of Chemical Physics*. 2000, *113*(10), 4139–4145.

[91] Varandas, A. J. C., Zhang, L Test Studies on the Potential Energy Surface and Rate Constant for the $OH + O_3$ Atmospheric Reaction. *Chemical Physics Letters*. 2000, *331*(5–6), 474–482.

[92] *Atmosfera. Spravochnik* [*Atmosphere: A Handbook*]; Gidrometeoizdat: Leningrad, 1991, 510 pp.

[93] Okabe, H. *Photochemistry of Small Molecules;* Wiley: New York, 1978, 431 pp.

[94] Pikaev, A. K. *Sovremennaya radiatsionnaya khimiya. Radioliz gazov i zhidkostei* [*Modern Radiation Chemistry: Radiolysis of Gases and Liquids*]; Nauka: Moscow, 1986, 440 pp.

[95] Boyd, A. W., Willis, C., Miller, O. A. A Re-examination of the Yields in the High Dose Rate Radiolysis of Gaseous Ammonia. *Canadian Journal of Chemistry*. 1971, *49*(13), 2283–2289.

[96] Silaev, M. M. Competition Mechanism of Substrate-Inhibited Radical Chain Addition to Double Bond. *Neftekhimiya*. 2000, *40*(1), 33–40 [English Translation in: *Peroleum Chemistry*. 2000, *40*(1), 29–35].

[97] Silaev, M. M. Competition Kinetics of Nonbranched Chain Processes of Free Radical Addition to the C=C, C=O, and O=O Double Bonds of Molecules. *Neftekhimiya*. 2003, *43*(4), 302–307 [English Translation in: *Petroleum Chemistry*. 2003, *43*(4), 258–273].

[98] Silaev, M. M. Low-reactive Free Radicals Inhibiting Nonbranched Chain Processes of Addition. *Biofizika*. 2005, *50*(4), 585–600 [English Translation in: *Biophysics*. 2005, *50*(4), 511–524].

[99] Sanderson, R. T. Radical Reorganization and Bond Energies in Organic Molecules. *The Journal of Organic Chemistry*. 1982, *47*(20), 3835–3839.

[100] Vereshchinskii, I. V., Pikaev, A. K. *Vvedenie v radiatsionnuyu khimiyu* (*Introduction to Radiation Chemistry*); Spitsyn, V.I., Ed., Akademiya Nauk SSSR: Moscow, 1963, p. 190.

Appendix

Table. The experimental concentrations x (mol dm^{-3}) of free formaldehyde at different temperatures T (K) and total formaldehyde concentrations c_0 (mol dm^{-3}) in various solvents

c_0	T	10^2x	c_0	T	10^2x	c_0	T	10^2x	c_0	T	10^2x
Water			4.44	389	5.20	4.0	381	5.00	1.8	371	2.08
1.0	358	0.78	4.44	405	7.50	4.0	397	8.80	1.8	393	6.00
1.0	387	2.22	4.44	418	10.0	4.0	409	12.00	1.8	418	12.20
1.0	393	3.23	Methanol			6.2	347	2.80	1.8	438	16.70
1.0	407	4.55	1.0	375	0.33	6.2	376	7.80	3.0	343	1.25
2.0	353	1.44	1.0	395	1.00	6.2	393	12.50	3.0	375	5.40
c_0	T	10^2x	c_0	T	10^2x	c_0	T	10^2x	c_0	T	10^2x
2.0	387	4.70	1.0	423	2.90	1-Propanol			3.0	403	15.80
2.0	397	6.60	2.5	373	0.60	1.0	371	0.83	3.0	413	19.40
2.0	407	8.55	2.5	385	1.15	1.0	393	2.10	5.6	343	2.80
4.0	343	0.78	2.5	398	1.80	1.0	413	4.30	5.6	358	3.35
4.0	363	2.33	5.4	351	0.78	1.0	435	7.65	5.6	363	5.80
4.0	385	6.45	5.4	383	3.70	1.9	353	0.70	5.6	371	6.50
4.0	403	8.90	5.4	398	6.80	1.9	383	3.06	5.6	383	12.10
4.0	413	11.10	7.0	365	4.70	1.9	405	7.65	2-Methyl-2-propanol		
6.0	351	2.22	7.0	383	12.50	1.9	417	11.70	1.0	347	1.20
6.0	375	6.70	7.0	391	16.00	4.0	349	1.67	1.0	367	4.50
6.0	389	10.70	Ethanol			4.0	373	6.10	1.0	387	11.00
6.0	398	14.10	1.0	367	0.33	4.0	393	13.30	1.0	398	19.30
8.4	364	5.50	1.0	387	0.67	6.0	338	1.39	2.0	335	1.10
8.4	376	8.32	1.0	397	1.45	6.0	357	5.00	2.0	357	4.30
8.4	388	10.97	1.0	413	2.70	6.0	377	11.70	2.0	375	13.00
Ethanediol			1.0	423	4.00	6.0	389	18.30	2.0	383	18.50
1.0	409	1.30	2.0	373	1.10	7.8	343	3.06	3.0	338	1.70
1.0	418	1.80	2.0	394	2.90	7.8	358	6.25	3.0	353	4.70
c_0	T	10^2x	c_0	T	10^2x	c_0	T	10^2x	c_0	T	10^2x
1.0	435	2.45	2.0	409	5.80	7.8	377	16.90	3.0	365	9.60
3.33	358	1.20	2.0	419	8.20	2-Propanol			3.0	373	15.50
3.33	387	3.30	3.0	361	1.20	1.0	365	0.98	6.0	345	6.90
3.33	401	5.10	3.0	387	3.70	1.0	393	3.05	6.0	351	9.00
3.33	415	7.20	3.0	409	7.80	1.0	411	6.00	6.0	361	13.40
4.44	338	1.00	4.0	355	2.30	1.0	433	10.40	6.0	365	18.30

Chapter 4

Kinetic Study and Mechanistic Elucidation of Metformin Oxidation by Mn(VII) in Alkaline Medium

Rajesh N. Hegde* **and P. Vishwanatha**
Department of Chemistry,
Sri Dharmasthala Manjunatheshwara College (Autonomous), Ujire, India

Abstract

The kinetics of the oxidation of an anti-diabetic drug, metformin using permanganate in alkaline medium was studied spectrophotometrically by keeping 0.01 mol dm^{-3} ionic strength. The oxidation of metformin by permanganate in NaOH media shows 1:6 stoichiometry (metformin: permanganate). The oxidation reaction is first order in [permanganate ion] and less than unit order in metformin and hydroxyl ion, respectively. When ionic strength of reaction medium increase, also there is an increment in reaction rate. The consequences of the variation of the dielectric constant of the media and the addition of external products were also studied. A suitable mechanism for the oxidation was proposed. The main product of the above mentioned reaction was recognized using liquid chromatography-mass spectrometry technique. The reaction constants corresponding to various steps of the mechanism were computed. Moreover, the activation constants and several thermodynamic values are also evaluated by graphical methods and also discussed in detail.

Keywords: oxidation, metformin, kinetics, permanganate

---

* Corresponding Author's Email: rajeshhegd@gmail.com.

In: A Closer Look at Chemical Kinetics
Editor: Victor Martinez-Luaces
ISBN: 979-8-88697-484-3

Introduction

Diabetes, the fastest growing disease in world, is categorized as insulin dependent (type 1) and not insulin dependent (type 2). Metformin (N,N-dimethylbiguanide) (MET), is a biguanide prescribed and employed in treatment of type 2 diabetes[1] and is identified to raise glucose transfer crossways in the cell membrane in skeletal muscles. This drug was the main chosen drug for overweight patients [2] and also diminishes the creation of advanced glycosylation end products [3]. According to International Diabetes Federation, one in ten people are living with diabetes. The statistical data reveals that, adults having diabetes are about 463 million worldwide and 700 million adults will have diabetes with in 2045, the prediction says. The diabetic people are more vulnerable to severe effect of corona virus. MET has common adverse effects such as diarrhea, nausea and abdominal pain and is not recommended for those with significant liver disease [4].

Permanganate ion was used for the oxidation of various substrates and it also had a broad range of relevance in organic synthesis also [5-9], mainly in phase transfer catalysis which uses the solvents like benzene and methylene chloride [7-9]. The reaction medium, which is used to carry out the reaction has much control over the oxidation mechanism of a substrate by permanganate ion [10]. In a strong alkaline medium, the established reduction product of MnO_4^- was MnO_4^{2-}. Because of lack of mechanistic proof, it is not possible to differentiate between a direct one-electron reduction to Mn(VI) and a reaction wherein two electron step involved to form Mn(V) followed by a fast reaction [11-13]. In sight of the popularity and medicinal significance of MET and short of evidence about the oxidation reaction mechanism of MET by Mn(VII), it was needed to know the mechanism of oxidation this drug molecule. The present investigation may help to obtain some information of the drug in biological system. Therefore the title reaction was studied carefully in basic medium using Mn(VII). The intention of the current study is (1) to obtain a suitable rate law for the oxidation reaction (2) suggest a proper mechanism for the reaction (3) recognize the species involved in the reaction, and (4) know the products of oxidation of MET.

Experiments

Doubly distilled water and analytical grade chemicals were employed during the job. Hence, by dissolving a proper amount of MET (Alfa Aser) in doubly distilled water, the stock solution was prepared. From the stock solution the required amount of MET was used. The strength of prepared potassium permanganate (Nice Chem Ltd) solution was determined using oxalic acid [14]. As documented in literature [15], the solution of potassium manganate was prepared. The potassium permanganate solution was boiled at about 100°C with KOH solution (8.0 mol dm^{-3}) until the color changed to green. The potassium manganate solid obtained by cooling the above content and using KOH solution it was recrystallized. Using this sample, potassium manganate stock solution was prepared using aqueous solution of KOH. In order to standardize the prepared solution, its absorbance was noted at 607 nm (ε = 1520 ± 30 dm^3 mol^{-1} cm^{-1}). The solutions of NaOH and $NaClO_4$ were prepared by doubly distilled water and were used to give the essential alkalinity and retain the ionic strength, respectively. Stability studies conducted indicated that the alkali had no effect on the constancy of MET in the absence of permanganate in the present experimental conditions.

A Bio UV-vis spectrophotometer (Varian CARY 50, Varian, Victoria-3170, Australia) connected to an accessory (peltier accessory for temperature control) was employed for the kinetic measurements. A Liquid Chromatography (Waters, Synapt G2, USA) connected to a mass spectrometer having electron ionization (EI) system, was employed to identify products. A pH meter was utilized to measure pH.

At a temperature of 25 ± 0.2°C and maintaining pseudo-first-order conditions, the entire kinetic measurements were monitored by taking surplus quantity of MET over the quantity of permanganate by maintaining a steady ionic strength. All the solutions were thermostated and the reaction was begun by adding permanganate and MET, which also has necessary amounts of $NaClO_4$ and NaOH for maintaining necessary ionic strength and alkalinity, respectively. The path of this reaction was monitored by measuring the reduction in optical density of Mn(VII) at its maximum optical density at 525 nm. The validity of Beer's law at 525 nm to Mn(VII) was checked, obtaining ε = 2096 ± 40 dm^3 mol^{-1} cm^{-1} (lit. ε = 2200 dm^3 mol^{-1} cm^{-1}). Additionally, using the graph of $\log_{10}(A_t\text{-}A_\infty)$ against time the rate constants of first order, k_{obs} were computed by substituting the data into the formula $A_t = A_\infty + (A_0 - A_\infty)\exp(-k_{obs}t)$, where A_t, A_0, and A_∞ are absorbance of Mn(VII) at time t, 0, and ∞, respectively. The obtained graphs were more or less straight lines up to 80% completion of oxidation, and the errors in k_{obs} values were about ±5% limit. While monitoring the title reaction, reaction mixture colour altered to blue from violet and later was green. A later obtained UV-Visible band of green solution is matching with that of MnO_4^{2-}. This was likely that, blue colour derived from the violet of Mn(VII) and the manganate shows green colour, without growing of hypomanganate {Mn(V)}. It was also observed while following the reaction that, the absorbance increases at 608 nm and a decline in optical density at 525 nm, which indicated the transformation of Mn(VII) to Mn(VI).

Results and Discussion

In the presence of steady quantities of NaOH and sodium perchlorate, dissimilar sets of reaction blends having uneven proportions of Mn(VII) to MET were kept in a sealed vessel underneath an inert condition for 3 hours. Then, the left behind quantity of Mn(VII) was projected spectrophotometrically at 525 nm. The outcome pointed out a 1:6 stoichiometry as shown in Equation 1.

(1)

The reaction solution of permanganate and MET having necessary amount of NaOH was made to react for 32 hours at 25°C. Using a hydrochloric acid solution, the product of the reaction was neutralized. Utilizing ethyl acetate solvent, the aqueous layer present was extracted four times, and this was used for thin layer chromatography. The main product of oxidation of MET, N-(N,N-dimethyl carbamimidoyl)-1-nitro formimidamide, was established by Liquid Chromatographic-Mass Spectroscopic analysis. A peak at 160.13 amu corresponding to molecular ion (Figure 1) confirms the existence of the product of the reaction, N-(N,N-dimethyl carbamimidoyl)-1-nitro formimidamide. It was observed that in the present reaction situation, N-(N,N-dimethyl carbamimidoyl)-1-nitro formimidamide was not undergoing further oxidation reaction.

Moreover, by fluctuating the concentration of OH^- and MET and by maintaining all additional quantities and circumstances steady, the order of the reaction was computed by a graphical method using the slopes of the graphs of $\log_{10} k_{obs}$ against $\log_{10}$ (concentration).

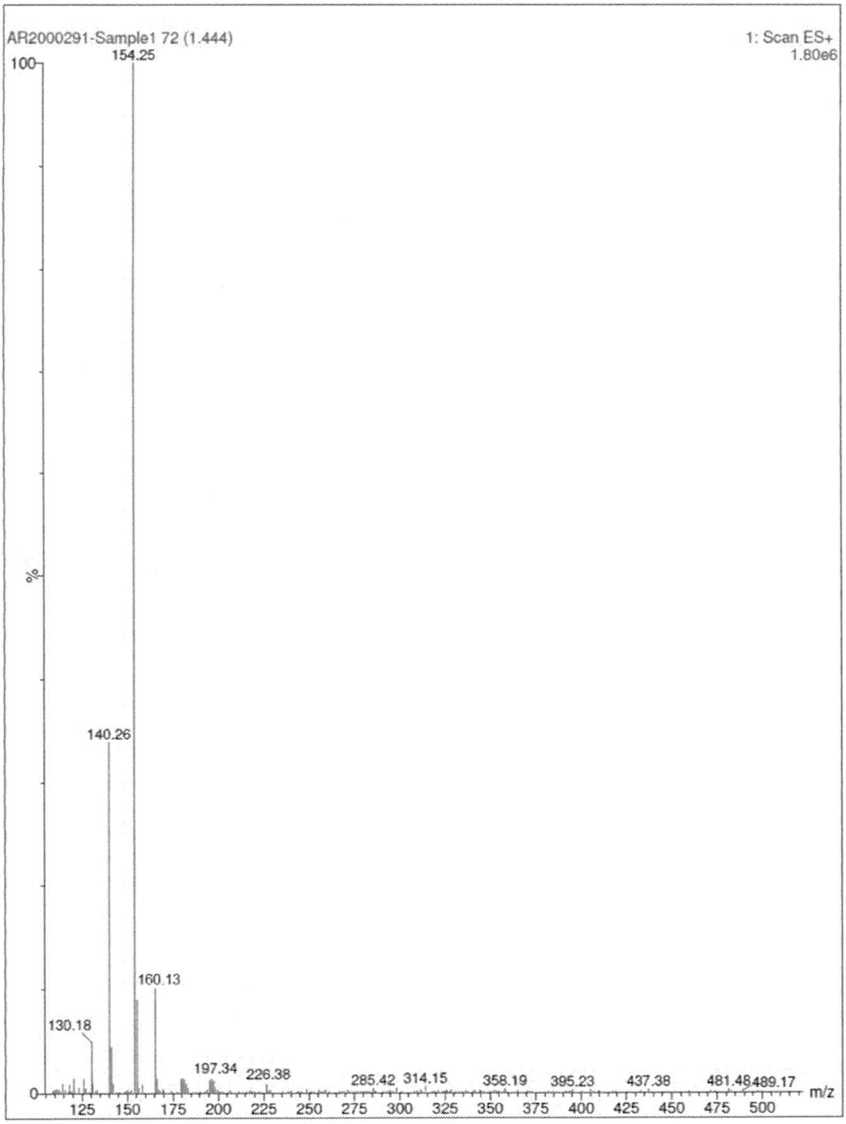

Figure 1. LC-MS spectra of N-(N,N-dimethyl carbamimidoyl)-1-nitro formimidamide (M+1) m/z at 160.13 amu.

Table 1. Influence of $[MnO_4^-]$, [MET] and $[OH^-]$ on the oxidation of metformin by permanganate in alkaline medium at 25°C

$[MnO_4^-]$ x10^4 (mol dm^{-3})	[MET] x10^3 (mol dm^{-3})	$[OH^-]$ x10^2 (mol dm^{-3})	k_{obs} x10^2 (s^{-1})	k_{cal} x10^2 (s^{-1})
0.2	1.0	4.0	2.7	2.8
0.4	1.0	4.0	2.9	2.8
0.6	1.0	4.0	2.7	2.8
0.8	1.0	4.0	2.8	2.8
1.0	1.0	4.0	2.8	2.8
2.0	1.0	4.0	2.8	2.8
1.0	0.4	4.0	1.4	1.5
1.0	0.6	4.0	2.0	2.0
1.0	0.8	4.0	2.4	2.4
1.0	1.0	4.0	2.8	2.8
1.0	2.0	4.0	3.8	3.9
1.0	4.0	4.0	5.2	5.0
1.0	1.0	0.6	1.7	1.7
1.0	1.0	0.8	1.9	1.9
1.0	1.0	1.0	2.1	2.1
1.0	1.0	2.0	2.4	2.5
1.0	1.0	4.0	2.7	2.8
1.0	1.0	6.0	3.0	2.9

In addition, at a fixed concentration of MET, 0.001 mol dm^{-3}, and OH^-, 0.04 mol dm^{-3} at a steady 10.0 x 10^{-2} mol dm^{-3} ionic strength, the oxidant quantity was fluctuated between 0.00002 to 0.000002 mol dm^{-3}. As seen in Table 1, fairly constant values of k_{obs} points out that, the order in [Mn(VII)] was one. The graph of $\log_{10} [A_t - A_\infty]$ against time ($r^2 \geq 0.9852$) was linear to about 80% finishing of the oxidation reaction (Figure 2), points out that the oxidant order was one. The last column in Table 1, k_{cal} values were matching with k_{obs} values.

The influence of quantity of the substrate on the title reaction has been examined by maintaining all other quantities and situations constant, the substrate quantity was assorted from 0.0004 to 0.00004 mol dm^{-3} at 25°C. With increase in quantity of the substrate MET, the k_{obs} values also increased (Table 1). At these reaction environment and concentrations, the order in [MET] was determined to be less than unity. However, at elevated quantities of the substrate, the oxidation reaction was not dependent on [MET] and at lower quantities of MET, it was found that the reaction was first order in [MET].

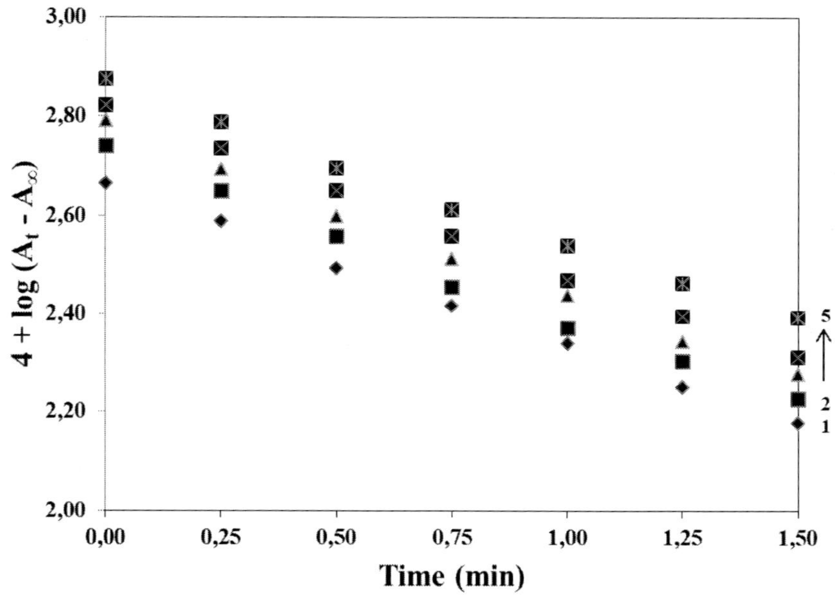

Figure 2. Graph showing first order dependence of [Mn(VII)] during the oxidation of metformin by Mn(VII) in alkaline media at 25°C $[MnO_4^-] \times 10^{-4}$ (mol dm^{-3}): (1) 0.2, (2) 0.4, (3) 0.6, (4) 1.0, (5) 2.0, ([MET] = 1.0×10^{-3}, [NaOH] = 0.04 and Ionic strength = 0.10 mol dm^{-3} respectively).

The influence of hydroxyl ion concentration was evaluated by maintaining the oxidant, ionic strength and substrate quantity study. The rate of the reaction raised as alkali concentration increased as can be observed in Table 1, and order with hydroxyl ion concentration resulted smaller than one. The order alters to zero order from the first order with increasing concentration of alkali.

The consequence of ionic strength on the title reaction was examined by altering the quantity of sodium perchlorate. The perchlorate was altered from 0.10 to 0.50 mol dm^{-3} by holding other situations and quantities steady. The study showed an increment of the k_{obs} values increased as sodium perchlorate concentration increases. A graph of log k_{obs} against $I^{1/2}$ is a straight line (solid line in Figure 3) with a positive slope. Dielectric constant of the reaction media has been altered using dissimilar quantities of tert-butyl alcohol and water.

Thus by diminishing the dielectric constant of the reaction medium, it is observed a reduction in reaction rate, and a graph of log k_{obs} against $1/D$ is a straight line with a negative slope (dashed line in Figure 3).

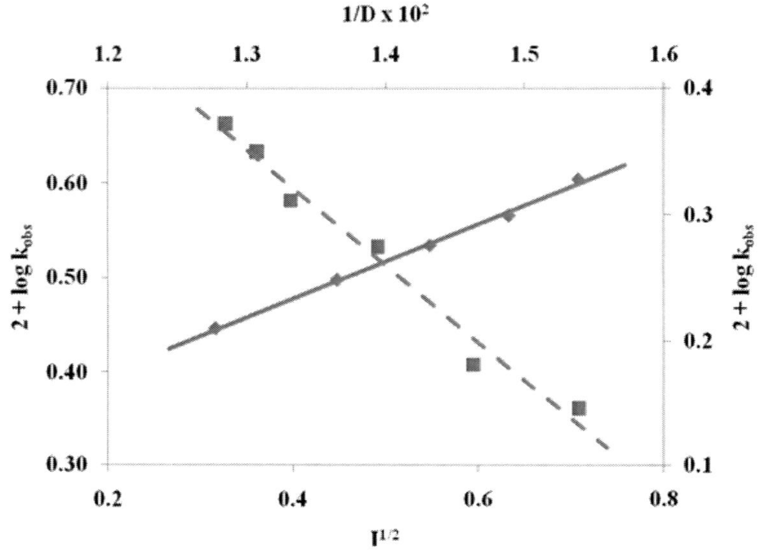

Figure 3. Influence of ionic strength (solid line) and dielectric constant (dashed line) of the reaction media on the oxidation of metformin by Mn(VII) at 298 K.

It was found that the rate of the reaction is not significantly altered by adding the product in the form of Mn(VI) at the beginning. The title reaction was followed in a sealed vessel containing nitrogen to study rate of the reaction. Substantial deviations in the results are not observed when the air atmosphere is changed by a nitrogen one.

The effect of carbonate on rate of reaction has been studied. It was observed that, reaction rates do not show any deviation by added carbonate.

The rate law can be written based on the above experimental observations as:

$$\text{Rate} = \frac{d[MnO_4^-]}{dt} = k_{obs}[MnO_4^-][MET]^{0.63}[OH^-]^{0.22}$$

To examine the free radicals involvement in the title reaction, a known amount of the monomer viz, acrylonitrile was introduced and kept aside for 2 hours in a sealed vessel with nitrogen. After stipulated time, a white precipitate was methanol was formed after adding methanol to the reaction mixture which indicates the involvement of free radicals in the reaction. However, a blank determination, consisting in runs of alone oxidant or MET with the monomer did not showed any white precipitation using the environment employed earlier. Although, it was observed that the reaction rates decreased when acrylonitrile was added, indicating the presence of free radicals in the reaction path as reported [12].

Besides, to analyze the effect of temperature on the title oxidation reaction, the measurements were done on four dissimilar temperatures using unvarying extents of MnO_4^-, MET, NaOH and $NaClO_4$. Utilizing the intercepts of the graph of $1/k_{obs}$ against 1/[MET] ($r^2 \geq 0.995$), the rate constants for the slow step in the Scheme 1 are calculated at four dissimilar temperatures. Employing the Arrhenius graph of $\log_{10} k$ against 1/T, the activation energy was obtained and from that value, the activation parameters were evaluated and were listed in Table 2.

Table 2. The values of K_1, K_2, and k at dissimilar temperatures for the oxidation of metformin by Mn(VII) in alkaline media

Temp (K)	$K_1 \times 10^{-2}$ (dm^3mol^{-1})	$K_2 \times 10^{-2}$ (dm^3mol^{-1})	$k \times 10^2$ (s^{-1})
288	1.22	9.48	15.02
293	1.15	8.68	21.32
298	1.09	8.49	27.94
303	1.03	8.27	40.92

In aqueous alkaline and acidic media, MnO_4^- is a powerful oxidant. Under the reaction conditions employed in the title reaction, the reduction product Mn(VI) was steady and additional reduction is not predicted, since the pH value is above 12 [15].

The studies conducted above pH 12 using diode array rapid scan spectrophotometric studies pointed out that, Mn(VII) was reduced to Mn(VI) and additional reduction did not take place. However, Mn(VI) is gradually reduced to Mn(IV), if the reaction time is extended. It was noticed that, the colour changes from violet of Mn(VII) ion to dark green of Mn(VI) ion via blue of Mn(V). The reaction between MET and Mn(VII) in strong alkaline media has 1:6 stoichiometry with fractional order in [MET] and in [permanganate], and the rate of reaction enhances with increasing the quantity of alkali. Products added does not show any influence on rate of reaction.

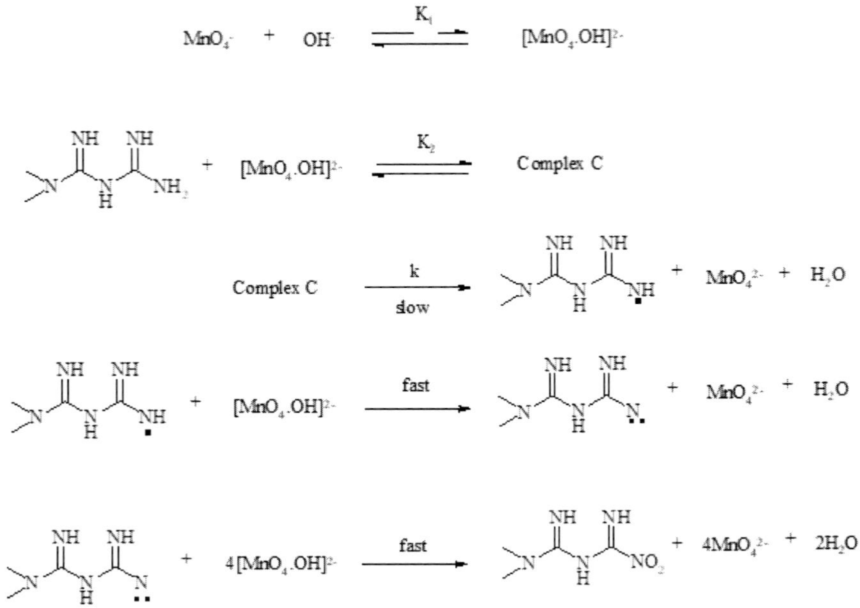

Scheme 1. The probable mechanism for oxidation reaction.

The mechanism in Scheme 1 was proposed for oxidation reaction of MET, taking into account all the orders observed experimentally. The results point out that there is a combination of OH^- with permanganate in the preceding equilibrium step to form an alkaline permanganate species [16-18], $[MnO_4.OH]^{2-}$. So formed permanganate species reacts with MET and gives Complex C. In a rate determining or slow step, Complex C decays and gives a MET free radical, MnO_4^{2-} and a molecule of water, which is followed by successive quick steps to give away the products of the reaction as proposed in Scheme 1.

Moreover, by analyzing the UV-Vis spectra of permanganate (1.0 x 10^{-4} mol dm^{-3}), $[OH^-]$ (0.04 mol dm^{-3}), and MET (1.0 x 10^{-3} mol dm^{-3}) and a blend of them, the creation of a complex between the oxidant, Mn(VII) and the substrate, (MET) can be proved. In fact, a shift of 6.0 nm is observed and this sort of complex formation was documented elsewhere [18]. With enhance in the amount of alkali, rate of reaction increases, which indicates the $[MnO_4.OH]^{2-}$ formation [18]. In agreement with the well-established principle of non-complimentary oxidation reactions which proceeds in a succession of one electron steps, an oxidant and a substrate would have a radical intermediate in the reaction as depicted in Scheme 1. The positive test for polymerization showed the presence of short-lived free radicals. The rate law corresponding to Scheme 1 is:

$$\text{Rate} = \frac{d[MnO_4^-]}{dt} = k\,[C] \tag{2}$$

$$= kK_1K_2[MET]_f[MnO_4^-]_f[OH^-]_f \tag{3}$$

The total $[MnO_4^-]$ is given by:

$$[MnO_4^-]_T = [MnO_4^-]_f + [MnO_4.OH]^{2-} + [C]$$

$$= [MnO_4^-]_f\ \{1 + K_1[OH^-] + K_1K_2[MET][OH^-]\} \tag{4}$$

The terms $[MnO_4^-]_T$ and $[MnO_4^-]_f$ depict the MnO_4^- concentrations, being T and f the total and free concentrations, respectively. So, free $[MnO_4^-]$ can be written as:

$$[MnO_4^-]_f = \frac{[MnO_4^-]_T}{1 + K_1[OH^-] + K_1K_2[MET][OH^-]} \tag{5}$$

Likewise $[OH^-]$ was calculated as:

$$[OH^-]_f = \frac{[OH^-]_T}{1 + K_1[MnO_4^-] + K_1K_2[MET][MnO_4^-]}$$

But concentration of MnO_4^- and MET employed in this experiment were low and the denominator in the formula approximate to one. Hence,

$$[OH^-]_f \cong [OH^-]_T \tag{6}$$

Likewise,

$$[MET]_f \cong [MET]_T \tag{7}$$

Equation 3, after substituting Equations 5, 6, and 7, and excluding T and f, takes the following form:

$$\text{Rate} = \frac{kK_1K_2[MET][MnO_4^-][OH^-]}{1 + K_1[OH^-] + K_1K_2[MET][OH^-]} \tag{8}$$

or,

$$\frac{\text{Rate}}{[MnO_4^-]} = k_{obs} = \frac{kK_1K_2[MET][OH^-]}{1 + K_1[OH^-] + K_1K_2[MET][OH^-]} \tag{9}$$

The above Equation (9) explains all the orders observed experimentally in diverse species involved in the reaction. This Equation (9) was rearranged to equation (10), which was appropriate for the rate law verification.

$$\frac{1}{k_{obs}} = \frac{1}{kK_1K_2[MET][OH^-]} + \frac{1}{K_1K_2[MET]} + \frac{1}{k} \tag{10}$$

In accordance with the above equation, the graphs of $1/k_{obs}$ against $1/[MET]$ ($r^2 \geq 0.9958$) and $1/k_{obs}$ against $1/[OH^-]$ ($r^2 \geq 0.9974$) must be straight lines with intercepts which validates the creation of $[MnO_4.OH]^{2-}$ - MET complex, and this was confirmed at dissimilar temperatures (Figure 4). The k, K_1, and K_2 values at 25°C were computed as 27.94 x 10^{-2} s^{-1}, 1.09 x 10^2 dm^3 mol^{-1}, and 8.49 x 10^2 dm^3 mol^{-1}, respectively using slopes and intercepts of these graphs. The value of K_1 is in agreement with the reported data [18]. These values were used to compute the rate constants at dissimilar reaction situations using Equation (9) and are compared with experimental results, supporting the proposed Scheme 1.

As depicted in Scheme 1, the reaction was among two species which were negatively charged was fortified, since reaction rate increases with a raise in ionic strength of the medium. In this study, the graph obtained has a negative slope as seen from Figure 3, which supports the fact that negative ions were involved in the reaction as shown in Scheme 1.

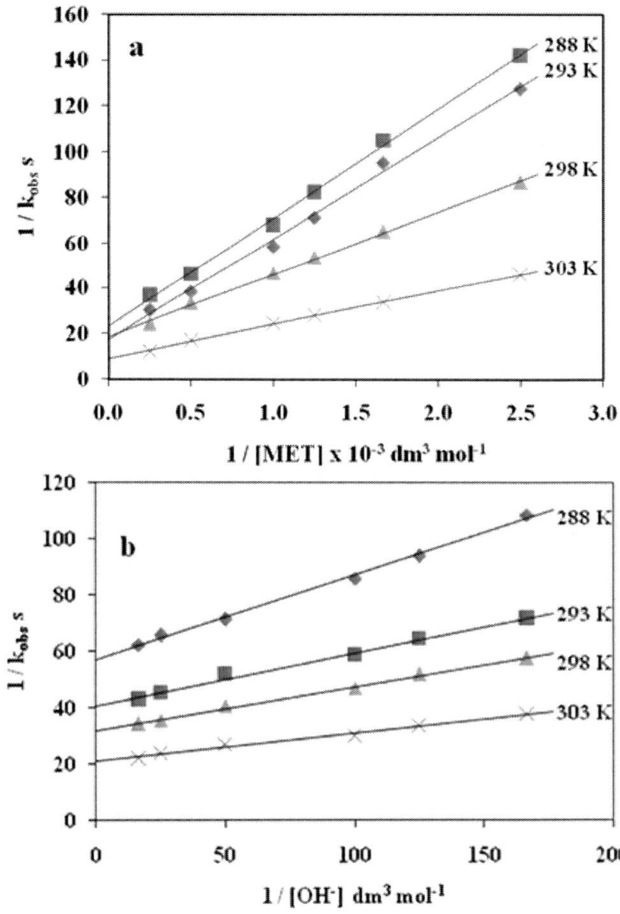

Figure 4. (a) Graphs of $1/k_{obs}$ versus 1/[MET] at four dissimilar temperatures (conditions as in Table 1). (b) Graphs of $1/k_{obs}$ versus 1/[OH^-] at four dissimilar temperatures (conditions as in Table 1).

Also, we computed quantities related to thermodynamics for the step first step equilibrium (Scheme 1). The [OH^-] and [MET] were varied (Table 1) at dissimilar temperatures. The values of K_1 and K_2 were obtained at four dissimilar temperatures and using slopes and intercepts of graphs of $1/k_{obs}$ against 1/[MET] ($r^2 \geq 0.9958$) and $1/k_{obs}$ against 1/[OH^-] ($r^2 \geq 0.9974$), and are listed in Table 2. The obtained values of the heat of the reaction (ΔH), the entropy (ΔS), and free energy of the reaction (ΔG), were computed from the van't Hoff's graph for deviation of K_1 with temperature; log K_1 against 1/T ($r^2 \geq 0.9986$) and are shown in Table 3.

Table 3. The activation parameters and thermodynamic quantities computed for the title reaction

(a)Activation parameters				
E_a (kJ mol^{-1})	$\Delta H^{\#}$ (kJ mol^{-1})	$\Delta S^{\#}$ (J K^{-1} mol^{-1})	$\Delta G^{\#}$ (kJ mol^{-1})	$\log_{10}A$
47 ± 2	45 ± 1	-123 ± 2	81 ± 2	7 ± 1
(b)Thermodynamic quantities				
	values from K_1		values from K_2	
ΔH (kJ mol^{-1})	-10.42 ± 1		-8.73 ± 1	
ΔS (J K^{-1} mol^{-1})	-314.52 ± 5		-308.86 ± 5	
ΔG_{298} (kJ mol^{-1})	83.31 ± 3		83.31 ± 4	

The values corresponding to the slow step are compared with latter values indicating that, they mainly chiefly depict the limiting rate step supporting that reaction was fast before the slow step and involve modest energy of activation [20]. Similarly, at four dissimilar temperatures, the values of K_2 were computed and thermodynamic quantities were obtained (Table 3). The value of $\Delta S^{\#}$ was negative which indicates that reactants are less ordered than the intermediate complex [21]. The moderate values of $\Delta H^{\#}$ and $\Delta S^{\#}$ correspond to a reaction with an electron transfer.

Conclusion

The oxidation of antidiabetic drug metformin with Mn(VII) in alkaline media was analysed kinetically. In the title reaction, vital role was played by the pH of the reaction environment. Since MnO_4^-, the oxidant species, needs a pH of more than 12 (or else the system would be distressed), the further reaction yields Mn(IV) as the reduced product. The depiction of reaction mechanism of title reaction is reliable with all observations made including kinetic, mechanistic, products and spectral observations. The thermodynamic quantities were determined for all steps in the mechanism. For the slow step of the reaction, the activation parameters were evaluated.

References

[1] Setter, S. M., Iltz, J. L., Thams, J., Campbell, R. K., (2003). Metformin hydrochloride in the treatment of type 2 diabetes mellitus: a clinical review with a focus on dual therapy. *Clin. Therap.*, 25(12), 2991-3026.

[2] Susan, Z. Y., Jack, A. Y., (2014). Long-term Drug Treatment for Obesity A Systematic and Clinical Review. *JAMA.*, 311(1), 74-86.

[3] Ruggiero-Lopez, D., Lecomte, M., Moinet, G., Patereau, G., Lagarde, M., Wiernsperger, N., (1999). Reaction of metformin with dicarbonyl compounds. Possible implication in the inhibition of advanced glycation end product formation. *Biochem. Pharmacol.*, 58(11), 1765-1773.

[4] Brackett, C. C., (2010). Clarifying metformin's role and risks in liver dysfunction. *Am. Pharm. Assoc.*, 50(3), 407-410.

[5] Banerji, K. K., (1988). Mechanism of the oxidation of organic sulphides by permanganate ion. *Tetrahedron*, 44. 2969-2975.

[6] Stewart, R., (1965). *Oxidation in Organic Chemistry*, New York: Part A, K. B. Wiberg Ed, Academic Press.

[7] Lee, D. G., (1980). *The Oxidation of Organic Compounds by Permanganate Ion and Hexavalent Chromium*, La Salle: IL.

[8] Simandi, L. I., (1993). *The Chemistry of Functional Groups*, Chichester: Wiley.

[9] Lee, D. G., Lee, E. J., Brown, K. C., (1987). *Phase Transfer Catalysis, New Chemistry, Catalysis and Applications, American Chemical Society*, Washington DC: ACS Symposium Series No. 326, p 82.

[10] Jaky, L. I., Schelly, M. Z. A., (1985). Kinetics and mechanism of the permanganate ion oxidation of sulfite in alkaline solutions. The nature of short-lived intermediates. *J. Am. Chem. Soc.*, 107, 4220-4224.

[11] Farokhi, S., Nandibewoor, S. T., (2016). Mechanistic study of oxidative transformation of hetero aromatic stimulant 1,3,7-trimethyl uric acid by potassium permanganate in perchloric acid medium-a kinetic study. *J. Ind. Chem. Soc.*, 93(4), 427-435.

[12] Pol, P. D., Mahesh, R. T., Nandibewoor, S. T., (2002). Free radical intervention in the oxidation of nicotinamide by alkaline permanganate - A Kinetic study. *J. Chem. Res.*, 11, 533-534.

[13] Bolattin, M. B., Nandibewoor, S. T., Chimatadar, S. A., (2017). Influence of micellar aggregates on oxidative degradation of β-cyclodextrin by alkaline permanganate: a kinetic and mechanistic study of inclusion complex. *J. Env. Chem. Eng.*, 7, 47-52.

[14] Jeffery, G. H., Bassett, J., Mendham, J., Denny, R. C., (1996). *Vogel's Text Book of Quantitative Chemical Analysis, Essex: 5th ed*, ELBS Longman, U. K.

[15] Carrington, A., Symons, M. C. R., (1956). Structure and Reactivity of Oxyanions of Transition Metals Part I. The Manganese Oxy-anions. *J. Chem. Soc.*, 3373–3376.

[16] Hegde, R. N., Shetti, N. P., Nandibewoor, S. T., (2009). Kinetic and mechanistic investigations of oxidation of pentoxifylline drug by alkaline permanganate. *Ind. Eng. Chem. Res.*, 48, 7025-7031.

[17] Panari, R. G., Chougale, R. B., Nandibewoor, S. T., (1998). Oxidation of mandelic acid by alkaline potassium permanganate – A Kinetic Study. *J. Phys. Org. Chem.*, 11, 448-454.

[18] Farokhi, S. A., Kini, A. K., Nandibewoor, S. T., (2002). Oxidation of L-hydroxy proline by aqueous alkaline permanganate. A Kinetic Study. *Inorg. Rect. Mech.*, 4, 67-76.

[19] Amis, E. S., (1966). *Solvent Effects on Reaction Rates and Mechanisms,* New York: Academic Press.

[20] Seregar, V. C., Hiremath, C. V., Nandibewoor, S. T., (2006). Palladium(II) catalysed oxidation of L-proline by heptavalent manganese in aqueous alkaline medium: a free radical intervention and decarboxylation. *Tran. Met. Chem.,* 31, 541–548.

[21] Rangappa, K. S., Raghavendra, M. P., Mahadevappa, D. S., Channegowda, D., (1998). Sodium *N*-chlorobenzenesulfonamide as a selective oxidant for hexosamines in alkaline medium: a kinetic and mechanistic study. *J. Org. Chem.*, 63, 531-536.

Chapter 5

A New Mechanism of Radical-Chain Oxidation

Michael M. Silaev*
Chemistry Faculty,
Lomonosov Moscow State University,
Vorobievy Gory, Moscow, Russia

Abstract

The reaction scheme proposed for the initiated nonbranched-chain addition of free radicals to the multiple bond of the molecular oxygen includes the addition reaction of the peroxyl free radical to the oxygen molecule to form the tetraoxyl free radical. This reaction competes with chain propagation reactions through a reactive free radical. The chain evolution stage in this scheme involves a few free radicals, one of which (tetraoxyl) is relatively low-reactive and inhibits the chain process by shortening the kinetic chain length. Based on the suggested scheme rate equations (containing one to three parameters to be determined directly) are deduced using quasi-steady-state treatment. These kinetic equations were used to describe the γ-induced nonbranched-chain processes of free-radical oxidation of liquid *o*-xylene at 373 K and hydrogen dissolved in water containing various amounts of oxygen at 296 K.

The ratios of rate constants of competing reactions and rate constants of addition reactions to the molecular oxygen are defined. In these processes the oxygen with the increase of its concentration begins to act as an oxidation autoinhibitor (or an antioxidant), and the rate of peroxide formation as a function of the dissolved oxygen concentration has a maximum. It is shown that a maximum in these curves arises from the competition between hydrocarbon (or hydrogen) molecules and dioxygen for reacting with the emerging peroxyl 1:1 adduct radical. From

* Corresponding Author's E-mail: mmsilaev@gmail.com; michsil@yandex.ru.

In: A Closer Look at Chemical Kinetics
Editor: Victor Martinez-Luaces
ISBN: 979-8-88697-484-3

the energetic standpoint possible nonchain pathways of the free-radical oxidation of hydrogen and the routes of ozone decay *via* the reaction with the hydroxyl free radical in the upper atmosphere (including the addition yielding the hydrotetraoxyl free radical, which can be an intermediate in the sequence of conversions of biologically hazardous UV radiation energy) were examined. The energetics of the key radical-molecule gas-phase reactions is considered.

Keywords: low-reactive radical, autoinhibitor, competition, thermochemical data, hydrogen, energy

Introduction

The kinetics of inhibition for nonbranched-chain processes of saturated free-radical addition to the C=C and C=O double bonds of alkene and formaldehyde molecules, respectively, by low-reactive free radicals that can experience delocalization of the unpaired *p*-electron was first considered in [1]. In these processes, a low-reactive free radical is formed in the reaction competing with chain propagation reactions through a reactive free radical. In the present work, the kinetics of inhibition by low-reactive tetraoxyl free radicals is considered for nonbranched-chain processes of the addition of a free radical or an atom to one of the two multiply bonded atoms of the oxygen molecule yielding a peroxyl free radical and thus initiating oxidation, which is the basic process of chemical evolution. The peroxyl free radical then abstracts the most labile atom from a molecule of the compound being oxidized or decomposes to turn into a molecule of an oxidation product.

The only reaction that can compete with these two reactions at the chain evolution stage is the addition of the peroxyl radical to the oxygen molecule (provided that the oxygen concentration is sufficiently high). This reaction yields the secondary tetraoxyl 1:2 adduct radical, which is the heaviest and the largest among the reactants. It is less reactive than the primary peroxyl 1:1 adduct radical and, as a consequence, does not participate in further chain propagation. At moderate temperatures, the reaction proceeds *via* a nonbranched-chain mechanism. The aim of this study was the mathematical simulation of the oxidation process autoinhibited by oxygen, when the dependence of the peroxide formation rate on the dissolved oxygen concentration has a maximum. The simulation was based on experimental data obtained for γ-radiation-induced addition reaction of hydrocarbon free radical

$R^{\bullet}$ (or hydrogen atom) to the molecular oxygen for which the initiation rate V_1 is known (taking into account that $V = GP$ and $V_1 = \varepsilon G(R^{\bullet})P$, where P is the dose rate, ε is the electron fraction of the solvent RH in the binary hydrocarbon–dioxygen system [2], and $G(R^{\bullet})$ is the initial yield of the chain-carrier free radical $R^{\bullet}$ – initiation yield [3, 4]).

Based on the reaction scheme suggested for the kinetic description of the addition process to oxygen, the kinetic equations with one to three parameters to be determined directly were derived. Reducing the number of unknown parameters in a kinetic equation will allow one to decrease the narrowness of the correlation of these parameters and avoid a sharp buildup of the statistical error in the nonlinear estimation of these parameters in the case of a limited number of experimental data points. The rate constant of the addition to oxygen, estimated as a kinetic parameter, can be compared to its reference value if the latter is known. This provides a clear criterion to validate the mathematical description against experimental data.

Addition of Hydrocarbon Free Radicals

Usually, the convex curve of the hydrocarbon autooxidation rate as a function of the partial pressure of molecular oxygen ascends up to some limit and then flattens out [5]. When this is the case, the oxidation kinetics is satisfactorily describable in terms of the conventional reaction scheme [5–10], which involves two types of free radicals. These are the hydrocarbon radical $R^{\bullet}$ (addend radical) and the addition product $RO_2^{\bullet}$ (1:1 adduct radical). However, the existing mechanisms are inapplicable to the cases in which the rate of initiated oxidation as a function of the oxygen concentration has a maximum (Figures 1, 2) [11, 12]. Such dependences can be described in terms of the competition kinetics of free-radical chain addition, whose reaction scheme involves not only the above two types of free radicals but also the $RO_4^{\bullet}$ radical (1:2 adduct) inhibiting the chain process [13, 14].

The chain evolution (propagation and inhibition) stage of Scheme 1 includes consecutive reaction pairs 2, 3; 2, 3a; 3a, 3b; parallel (competing) reaction pairs 3, 3a; 3, 3b; 3, 4; 3a, 4, and consecutive–parallel reactions 2, 4.

The decomposition of the initiator I in reaction 1 yields a reactive $R_0^{\bullet}$ radical, which turns into the ultimate product R_0H *via* reaction 1a, generating an alkyl radical $R^{\bullet}$, which participates in chain propagation. In reaction 2, the addition of the free radical $R^{\bullet}$ to the oxygen molecule yields a reactive alkylperoxyl 1:1 adduct radical $RO_2^{\bullet}$ [10], which possesses increased energy owing to the energy released upon the conversion of the O=O bond into the ordinary bond RO–O$^{\bullet}$ (for addition in the gas phase under standard conditions, this energy is 115–130 kJ mol^{-1} for C_1–C_4 alkyl radicals [6, 16, 17] and 73 kJ mol^{-1} for the allyl radical [17]).

Scheme 1

Nonbranched-chain oxidation of hydrocarbons.[1]

Chain initiation:

1. $I \xrightarrow{2k_1} 2R_0^{\bullet}$;

1a. $R_0^{\bullet} + RH \xrightarrow{k_{1a}} R_0H + R^{\bullet}$.

Chain propagation:

2. $R^{\bullet} + O_2 \xrightarrow{k_2} RO_2^{\bullet}$;

3. $RO_2^{\bullet} + RH \xrightarrow{k_3} RO_2H + R^{\bullet}$
(or $ROH + RO^{\bullet}$);

3a. $RO_2^{\bullet} \xrightarrow{k_{3a}} R'_{(-H)}HO + R''O^{\bullet}$
(or $R_{(-2H)}HO + {}^{\bullet}OH$);

3b. $R''O^{\bullet}(\text{or } RO^{\bullet}) + RH \xrightarrow{k_{3b}} R''OH\ (\text{or } ROH) + R^{\bullet}$
(or ${}^{\bullet}OH + RH \xrightarrow{k_{3b}} H_2O + R^{\bullet}$)

Inhibition:

4. $RO_2^{\bullet} + O_2 \xrightarrow{k_4} RO_4^{\bullet}$.

---

[1] The only difference between the kinetic model of oxidation and the kinetic model of the chain addition of 1-hydroxyalkyl radicals to the free (unsolvated) form of formaldehyde in nonmethanolic alcohol–formaldehyde systems [1, 15] is that in the former does not include the formation of the molecular 1:1 adduct *via* reaction 4.

Chain termination:

5. $2R^{\bullet} \xrightarrow{2k_5} RR$
(or $R_{(-2H)}H + RH$);

6. $R^{\bullet} + RO_4^{\bullet} \xrightarrow{k_6} RH + R_{(-2H)}HO + O_3$
(or: $ROH + R_{(-2H)}HO + O_2$,
$ROR + O_3$,
$RO_2R + O_2$);

7. $2\,RO_4^{\bullet} \xrightarrow{2k_7} RO_2R + 2O_3$.

Because of this, the adduct radical can decompose (reaction 3a) or react with some neighbor molecule (reaction 3 or 4) on the spot, without diffusing in the solution and, accordingly, without entering into any chain termination reaction. In reaction 3, the interaction between the radical adduct $RO_2^{\bullet}$ and the hydrocarbon molecule RH yields, *via* a chain mechanism, the alkyl hydroperoxide RO_2H (this reaction regenerates the chain carrier $R^{\bullet}$ and, under certain conditions, can be viewed as being reversible [6]) or the alcohol ROH (this is followed by the regeneration of $R^{\bullet}$ *via* reaction 3b). The latter (alternative) pathway of reaction 3 consists of four steps, namely, the breaking of old bonds and the formation of two new bonds in the reacting structures. In reaction 3a, the isomerization and decomposition of the alkylperoxyl radical adduct $RO_2^{\bullet}$ with O–O and C–O or C–H bond breaking take place [5, 9], yielding the carbonyl compound $R'_{(-H)}HO$ or $R_{(-2H)}HO$. Reaction 3b produces the alcohol R″OH or water and regenerates the free radical $R^{\bullet}$ (here, R′ and R″ are radicals having a smaller number of carbon atoms than R). As follows from the above scheme of the process, consecutive reactions 3a and 3b (whose rates are equal within the quasi-steady-state treatment), in which the highly reactive fragment, oxyl radical $R''O^{\bullet}$ (or $^{\bullet}OH$) forms and then disappears, respectively, can be represented as a single, combined bimolecular reaction 3a,b occurring in a "cage" of solvent molecules. Likewise, the alternative (parenthesized) pathways of reactions 3 and 3b, which involve the alkoxyl radical $RO^{\bullet}$, can formally be treated as having equal rates. For simple alkyl C_1–C_4 radicals R, the pathway of reaction 3 leading to the alkyl hydroperoxide RO_2H is endothermic (ΔH°_{298}= 30–80 kJ mol^{-1}) and the alternative pathway yielding the alcohol ROH is exothermic (ΔH°_{298}= –120 to –190 kJ mol^{-1}), while the parallel reaction 3a, which yields a carbonyl compound and the alkoxyl radical $R''O^{\bullet}$ or the hydroxyl radical $HO^{\bullet}$, is exothermic in both cases

(ΔH°_{298}= −80 to −130 kJ mol^{-1}), as also is reaction 3b (ΔH°_{298}= −10 to −120 kJ mol^{-1}), consecutive to reaction 3a, according to thermochemical data for the gas phase [6, 17, 18]. In reaction 4, which is competing with (parallel to) reactions 3 and 3a (chain propagation through the reactive radical R$^{\bullet}$), the resulting low-reactive radical that does not participate in further chain propagation and inhibits the chain process is supposed to be the alkyltetraoxyl 1:2 radical adduct[2,3] $RO_4^{\bullet}$, which has the largest weight and size. This radical is possibly stabilized by a weak intramolecular H···O hydrogen bond [25] shaping it into a six-membered cyclic structure[4] (seven-membered cyclic structure in the case of aromatic and certain branched acyclic hydrocarbons) [27, 28]:

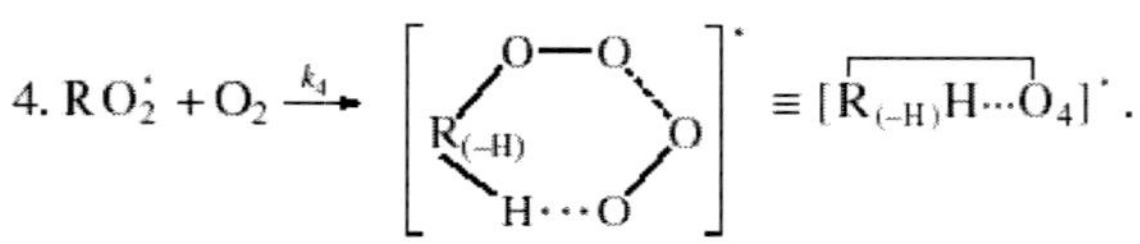

Reaction 4 in the case of the methylperoxyl radical $CH_3O_2^{\bullet}$ adding to the oxygen molecule to yield the methyltetraoxyl radical $CH_3O_4^{\bullet}$ takes place in the gas phase, with heat absorption equal to 110.0 ± 18.6 kJ mol^{-1} [20] (without the energy of the possible formation of a hydrogen bond taken into account). The exothermic reactions 6 and 7, in which the radical R$^{\bullet}$ or $RO_4^{\bullet}$ undergoes disproportionation, include the isomerization and decomposition of

---

[2] It is hypothesized that raising the oxygen concentration in the *o*-xylene–oxygen system can lead to the formation of an [ROO$^{\bullet}$···O_2] intermediate complex [11] similar to the [ROO$^{\bullet}$···(π-bond)RH] complex between the alkylperoxyl 1:1 adduct radical and an unsaturated hydrocarbon suggested in this work. The electronic structure of the π-complexes is considered elsewhere [19].

[3] Thermochemical data are available for some polyoxyl free radicals (the enthalpy of formation of the methyltetraoxyl radical without the energy of the possible intramolecular hydrogen bond H···O taken into account is $\Delta H_f{}^{\circ}{}_{298}(CH_3O_4^{\bullet})$ = 121.3 ± 15.3 kJ mol^{-1}) and polyoxides ($\Delta H_f{}^{\circ}{}_{298}(CH_3O_4H)$ = −21.0 ± 9 kJ mol^{-1}) [20]. These data were obtained using the group contribution approach. Some physicochemical and geometric parameters were calculated for the methyl hydrotetraoxide molecule as a model compound [21–23]. The IR spectra of dimethyl tetraoxide with isotopically labeled groups in Ar–O_2 matrices were also reported [24]. For reliable determination of the number of oxygen atoms in an oxygen-containing species, it is necessary to use IR and EPR spectroscopy in combination with the isotope tracer method [24].

[4] The $\overline{R_{(-H)}H\cdots O(R)O_3}$ ring consisting of the same six atoms (C, H, and 4O), presumably with a hydrogen bond [5], also forms in the transition state of the dimerization of alkylperoxyl radicals $RO_2^{\bullet}$ *via* the Russell mechanism [7, 26].

the $RO_4^•$ radical.[5] The latter process is likely accompanied by chemiluminescence typical of hydrocarbon oxidation [23]. These reactions regenerate oxygen as O_2 molecules (including singlet oxygen[6] [23, 30]) and, partially, as O_3 molecules and yield the carbonyl compound $R_{(-2H)}HO$ (possibly in the triplet excited state [23]). Depending on the decomposition pathway, the other possible products are the alcohol ROH, the ether ROR, and the alkyl peroxide RO_2R. It is likely that the isomerization and decomposition of the $RO_4^•$ radical *via* reactions 6 and 7 can take place through the breaking of a C–C bond to yield carbonyl compounds, alcohols, ethers, and organic peroxides containing fewer carbon atoms than the initial hydrocarbon, as in the case of the alkylperoxyl radical $RO_2^•$ in reaction 3a. At later stages of oxidation and sufficiently high temperatures, the resulting aldehydes can be further oxidized into respective carboxylic acids. They can also react with molecular oxygen so that a C–H bond in the aldehyde molecule breaks to yield two free radicals ($HO_2^•$ and $^•R'_{(-H)}O$ or $^•R_{(-2H)}O$). This process, like possible ozone decomposition yielding an $^•O^•$ atom or peroxide decomposition with O–O bond breaking, leads to degenerate chain branching [5].

The kinetic equations describing the rates of formation of molecular products at the chain propagation and termination stages of the above reaction Scheme 1 were obtained using the quasi-steady-state treatment. To simplify the writing of equations in the derivation process, we introduce the following notations: l = [RH] and x = [O_2]; $R_1 = R^•$, $R_2 = RO_4^•$, $R_3 = RO_2^•$, and $R_4 = R''O^•$ (or $^•OH$); $\alpha = k_3/k_4$ and $\beta = k_{3a}/k_4$ (mol dm^{-3}). According to Scheme 1, $V_1 = V_{1a}$, $V_{3a} = V_{3b}$, and $V_2 = V_3 + V_{3a} + V_4$. In order to reduce the exponent of the $2k_5[R^•]^2$ term in the $d[R^•]/dt = 0$ equation to unity [1], were used the following condition for the early stages of the process: $k_6 = \sqrt{2k_5 2k_7}$ [8] and, hence,

$$V_1 = V_5 + 2V_6 + V_7 = (\sqrt{2k_5}R_1 + \sqrt{2k_7}R_2)^2 \text{ and}$$

$$\sqrt{V_1} = (\sqrt{2k_5}R_1 + \sqrt{2k_7}R_2).$$

[5] Taking into account the principle of detailed balance for the various pathways of formation of products, whose numbers in the elementary reaction should not exceed three for possible involvement in the triple collisions in the case of the reverse reaction, since the probability of simultaneous interaction of four particles is negligible. This principle was not taken into account in the author's works on oxidation until 2012.

[6] Note that the alkylperoxyl radicals $RO_2^•$ are effective quenchers of singlet oxygen $O_2(a^1\Delta_g)$ [29].

Denote $\delta_1=\sqrt{2k_5V_1}$ and $\delta_2=\sqrt{2k_7V_1}$ and obtain the following expressions for free radicals.

$$d[\mathrm{R}_3]/dt = V_2 - V_3 - V_{3a} - V_4 = k_2\mathrm{R}_1x - k_3\mathrm{R}_3l - k_{3a}\mathrm{R}_3 - k_4\mathrm{R}_3x = k_2\mathrm{R}_1x - (k_3l + k_{3a} + k_4x)\mathrm{R}_3 = 0;$$

$$\mathrm{R}_3 = \frac{k_2\mathrm{R}_1x}{(k_3l + k_{3a} + k_4x)}.$$

$$d[\mathrm{R}_2]/dt = V_4 - 2V_6 - V_7 = k_4\mathrm{R}_3x - \delta_2\mathrm{R}_2 = \frac{k_2\mathrm{R}_1x^2}{(\alpha l + \beta + x)} - \delta_2\mathrm{R}_2 = k_2\mathrm{R}_1x^2 - (\alpha l + \beta + x)\delta_2\mathrm{R}_2 = 0;$$

$$\mathrm{R}_2 = \frac{k_2\mathrm{R}_1x^2}{(\alpha l + \beta + x)\delta_2}.$$

$$d[\mathrm{R}_1]/dt = V_1 - V_2 + V_3 + V_{3a} - V_5 - 2V_6 = V_1 - k_2\mathrm{R}_1x + k_3\mathrm{R}_3l + k_{3a}\mathrm{R}_3 - \delta_1\mathrm{R}_1 = V_1 - k_2\mathrm{R}_1x + \frac{k_2\mathrm{R}_1\alpha lx}{(\alpha l + \beta + x)} + \frac{k_2\mathrm{R}_1\beta x}{(\alpha l + \beta + x)} - \delta_1\mathrm{R}_1 = V_1(\alpha l + \beta + x) - [k_2x(\alpha l + \beta + x) - k_2\alpha lx - k_2\beta x + (\alpha l + \beta + x)\delta_1]\mathrm{R}_1 = V_1(\alpha l + \beta + x) - [k_2x^2 + (\alpha l + \beta + x)\delta_1]\mathrm{R}_1 = 0;$$

$$\mathrm{R}_1 = \frac{V_1(\alpha l + \beta + x)}{k_2x^2 + (\alpha l + \beta + x)\delta_1}.$$

$$V_3 = k_3\mathrm{R}_3l = \frac{k_2\mathrm{R}_1\alpha lx}{(\alpha l + \beta + x)} = \frac{V_1\alpha lk_2x}{k_2x^2 + (\alpha l + \beta + x)\sqrt{2k_5V_1}}.$$

$$V_{3a} = k_{3a}\mathrm{R}_3 = \frac{k_2\mathrm{R}_1\beta x}{(\alpha l + \beta + x)} = \frac{V_1\beta k_2x}{k_2x^2 + (\alpha l + \beta + x)\sqrt{2k_5V_1}}.$$

$V_5 = 2k_5\mathrm{R}_1^2$, $2V_6 = 2k_6\mathrm{R}_1\mathrm{R}_2$, and $V_7 = 2k_7\mathrm{R}_2^2$.

The rate constant k_2 of reaction 2 in the kinetic equations can be replaced with its analytical expression, which is obtained by solving the quadratic equation following from the rate function extremum condition $\partial V_{3,\,3a}(1{:}1\ \mathrm{Adduct})/\partial x = 0$.

$$V_{3,\,3a} = V_3 + V_{3a} = \frac{V_1(\alpha l+\beta)k_2x}{k_2x^2+(\alpha l+\beta+x)\delta_1} = \frac{\mathrm{D}x}{\mathrm{A}x^2+\mathrm{B}x+\mathrm{C}},$$

where $\mathrm{D} = V_1(\alpha l+\beta)k_2$, $\mathrm{A} = k_2$, $\mathrm{B} = \delta_1$, and $\mathrm{C} = (\alpha l+\beta)\delta_1$.

$$\partial V_{3,\,3a}/\partial x = \mathrm{D}(\mathrm{A}x^2+\mathrm{B}x+\mathrm{C}) - \mathrm{D}x(2\mathrm{A}x+\mathrm{B}) = -\mathrm{A}x^2+\mathrm{C} = -k_2x^2+(\alpha l+\beta)\delta_1 = 0.$$

The solution of this incomplete quadratic equation gives the value of the concentration x_m at the maximum point:

$$x_\mathrm{m} = \sqrt{\frac{(\alpha l_m+\beta)\delta_1}{k_2}}.$$

From here, the following kinetic equations of oxidation reaction rates were obtained.

$$V_3(\mathrm{RO_2H;\ ROH}) = V_1\alpha l k_2 x/f = \quad (1)$$

$$= V_1\alpha l x/f_m, \quad (1a)$$

$$V_{3a}(\mathrm{R'_{(-H)}HO;\ R_{(-2H)}HO}) = V_{3b}(\mathrm{R''OH;\ H_2O}) =$$

$$= V_1\beta\, k_2x/f = \quad (2)$$

$$= V_1\beta\, x/f_m, \quad (2a)$$

$$V_5 = V_1^2\,2k_5\,(\alpha l+\beta+x)^2/f^2, \quad (3)$$

$$2V_6 = 2V_1\sqrt{2k_5V_1}\,(\alpha l+\beta+x)k_2x^2/f^2, \quad (4)$$

$$V_7 = V_1\left(k_2x^2\right)^2 / f^2, \tag{5}$$

where V_1 is the initiation rate, l = [RH] and x = $[O_2]$ are the molar concentrations of the starting components ($l >> x$), $\alpha = k_3/k_4$ and $\beta = k_{3a}/k_4$ (mol dm^{-3}) are the ratios of the rate constants of the competing (parallel) reactions,

$$k_2 = (\alpha l_m + \beta)\sqrt{2k_5V_1}\,/\,x_m^2$$

is the rate constant of the addition of the alkyl radical R$^{\bullet}$ to the oxygen molecule (reaction 2) as determined by solving the quadratic equation following from the rate function extremum condition $\partial V_{3,\,3a} / \partial x = 0$, l_m and x_m are the values of l and x at the maximum point of the function, $f = k_2x^2 + (\alpha l + \beta + x)\sqrt{2k_5V_1}$ and $f_m = x^2 + (\alpha l + \beta + x)x_m^2/(\alpha l_m + \beta)$.

The ratios of the rates of the competing reactions are $V_3/V_4 = \alpha l/x$ and $V_{3a}/V_4 = \beta/x$, and the chain length is $n = (V_3 + V_{3a})/V_1$. The overall rate of the process is a complicated function of the formation and disappearance rates of the free radicals R$^{\bullet}$ and $RO_4^{\bullet}$: V(RO_2H, $R'_{(-H)}HO$, $R''OH$; ROH; $R_{(-2H)}HO$, H_2O) = $V_{1a} + V_3 + V_{3b} - V_4 - V_5 - V_7$. As distinct from the rates V_4 ($V_4 \leq V_1$), V_5, and V_7, the rates V_2, V_3, $V_{3a,\,3b}$, and $2V_6$ as a function of the concentration x have a maximum.

Equations (1a) and (2a) were obtained by replacing the rate constant k_2 in Equations (1) and (2) with its analytical expression (for reducing the number of unknown parameters to be determined directly).

For $\alpha l >> \beta$ ($V_3 >> V_{3a}$), when the total yield of alkyl hydroperoxides and alcohols having the same number of carbon atoms as the initial compound far exceeds the yield of carbonyl compounds, as in the case of the oxidation of some hydrocarbons, the parameter β in Equations (1) and (1a) can be neglected ($\beta = 0$):

$$f = k_2x^2 + (\alpha l + x)\sqrt{2k_5V_1},$$

$$f_m = x^2 + (\alpha l + x)x_m^2 / \alpha l_m \text{ and}$$

$$k_2 = \alpha l_m\sqrt{2k_5V_1}\,/\,x_m^2.$$

Equations (1) and (2) subject to the constraint $k_2x^2 >> (\alpha l + \beta + x)\sqrt{2k_5V_1}$ (the descending branch of the peaking curve) can be transformed into Equations (6) and (7), respectively, which express simple inverse proportionalities with respect to the oxygen concentration x and allow the parameters α and β to be tentatively estimated from the experimental product formation rate V provided that V_1 is known:

$$V_3 = V_1\alpha l/\varphi x\,, \tag{6}$$

$$V_{3a} = V_1\beta/\varphi x\,, \tag{7}$$

where $\varphi = 2$ at the point of maximum (where $k_2x^2 \cong (\alpha l + \beta + x)\sqrt{2k_5V_1}$) and $\varphi = 1$ for the descending branch of the curve.

In the alternative kinetic model of oxidation, whose chain termination stage involves, in place of $R^{\bullet}$ (Scheme 1), $RO_2^{\bullet}$ radicals reacting with one another and with $RO_4^{\bullet}$ radicals, the dependences of the chain formation rates of the products on the oxygen concentration x derived by the same method have no maximum:

$$V_3 = V_1k_3l/(k_4x + \sqrt{2k_{5_1}V_1}) \text{ and}$$

$$V_{3a} = V_1k_{3a}/(k_4x + \sqrt{2k_5V_1})\,.$$

In the kinetic model of oxidation that does not include the competing reaction 4 ($k_4 = 0$) and involves the radicals $R^{\bullet}$ and $RO_2^{\bullet}$ (the latter instead of $RO_4^{\bullet}$ in Scheme 1) in reactions 5–7, the reaction rate functions V_3 and V_{3a} obtained in the same way are fractional rational functions in the form of $a_0x/(b_0x + c_0)$, where a_0, b_0, and c_0 are coefficients having no extremum. For a similar kinetic model in which reactions 3a,b and 4 appearing in the above scheme are missing ($k_{3a} = k_4 = 0$), Walling [7], using the quasi-steady-state treatment in the long kinetic chain approximation, when it can be assumed that $V_2 = V_3$, without using the substitution $k_6 = \sqrt{2k_52k_7}$ [5, 7, 8] (as distinct from this work), found that $V_2 = V_3$ is an irrational function of: x: $a_1x/\sqrt{b_1x^2 + c_1x + d_1}$ where a_1, b_1, c_1, and d_1 are coefficients. Again, this function has no maximum with respect to the concentration of any of the two components.

Thus, of the three kinetic models of oxidation mathematically analyzed above, which involve the radicals $R^{\bullet}$ and in three types of quadratic-law chain termination reactions (reactions 5–7) and are variants of the conventional model [5–10], the last two lead to an oxidation rate versus oxygen concentration curve that emanates from the origin of coordinates, is convex upward, and has an asymptote parallel to the abscissa axis. Such monotonic dependences are observed when the oxygen solubility in the liquid is limited under given experimental conditions and the oxygen concentration attained[7] is $[O_2]_{top} \leq x_m$.

Unlike the conventional model, the above kinetic model of free-radical nonbranched-chain oxidation, which includes the pairs of competing reactions 3–4 and 3a–4 (Scheme 1), allows us to describe the nonmonotonic (peaking) dependence of the oxidation rate on the oxygen concentration (Figure 1). In this oxidation model, as the oxygen concentration in the binary system is increased, oxygen begins to act as an oxidation autoinhibitor or an antioxidant *via* the further oxidation of the alkylperoxyl 1:1 adduct radical $RO_2^{\bullet}$ into the low-reactive 1:2 adduct radical $RO_4^{\bullet}$ (reactions 4 and 6 lead to inefficient consumption of the free radicals $RO_2^{\bullet}$ and $R^{\bullet}$ and cause shortening of the kinetic chains). The optimum oxygen concentration x_m, at which the oxidation rate is the highest, can be calculated using kinetic equations (1a) and (2a) or the corresponding analytical expression for k_2. In the familiar monograph "Chain Reactions" by Semenov [31], it is noted that raising the oxygen concentration when it is already sufficient usually slows down the oxidation process by shortening the chains. The existence of the upper (second) ignition limit in oxidation is due to chain termination in the bulk through triple collisions between an active species of the chain reaction and two oxygen molecules (at sufficiently high oxygen partial pressures). In the gas phase at atmospheric pressure, the number of triple collisions is roughly estimated to be 10^3 times smaller than the number of binary collisions (and the probability of a reaction taking place depends on the specificity of the action of the third particle). Note that in the case of a gas-phase oxidation of hydrogen at low pressures of 25–77 Pa and a temperature of 77 K [12] when triple collisions

---

[7] The oxygen concentration attained in the liquid may be below the thermodynamically equilibrium oxygen concentration because of diffusion limitations hampering the establishment of the gas–liquid saturated solution equilibrium under given experimental conditions (for example, when the gas is bubbled through the liquid) or because the Henry law is violated for the given gas–liquid system under real conditions.

are unlikely, the dependence of the rate of hydrogen peroxide formation on oxygen concentration (the rate of passing of molecular oxygen *via* the reaction tube) also has a pronounced maximum (see curves *3* and *4* in Figure 2) that indicates a chemical mechanism providing the appearance of a maximum (see reaction 4 of Scheme 2).

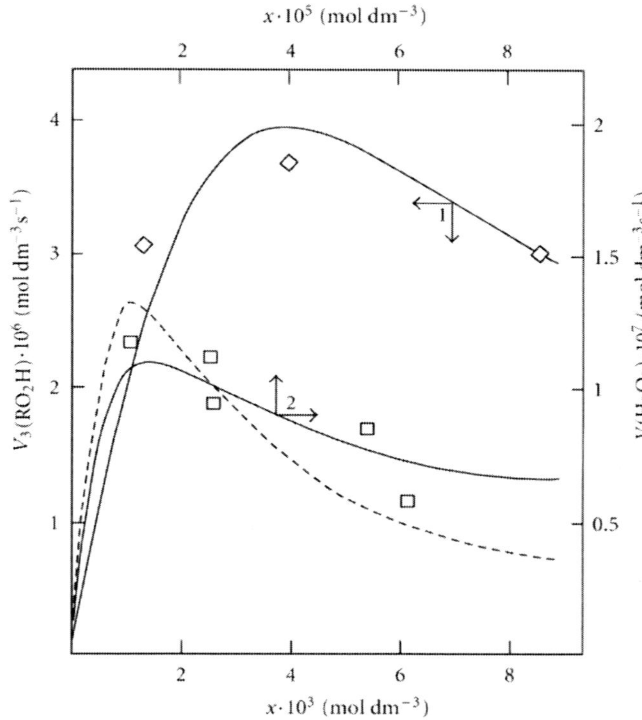

Figure 1. (*1*, ◇) Reconstruction of the functional dependence of the 2-methylbenzyl hydroperoxide formation rate $V_3(RO_2H)$ on the dissolved oxygen concentration x from empirical data (points) using Equation (1a) with $\beta = 0$ (model optimization with respect to the parameter α) for the *o*-xylene–oxygen system at 373 K [11] (standard deviation of $S_Y = 5.37 \times 10^{-7}$). (*2*, □) Reconstruction of the functional dependence of the total hydrogen peroxide formation rate $V_{3,7}(H_2O_2)$ on the dissolved oxygen concentration x from empirical data (symbols) using Equations (1a) and (5a) with $\beta = 0$ (model optimization with respect to the parameter α) for the γ-radiolysis of water saturated with hydrogen and containing different amounts of oxygen at 296 K [34] ($S_Y = 1.13 \times 10^{-8}$). The dashed curve described $V_3(H_2O_2)$ as a function of the oxygen concentration x based on Equation (1a) with $\beta = 0$ (model optimization with respect to α) and the experimental data of curve 2 ($S_Y = 1.73 \times 10^{-8}$).

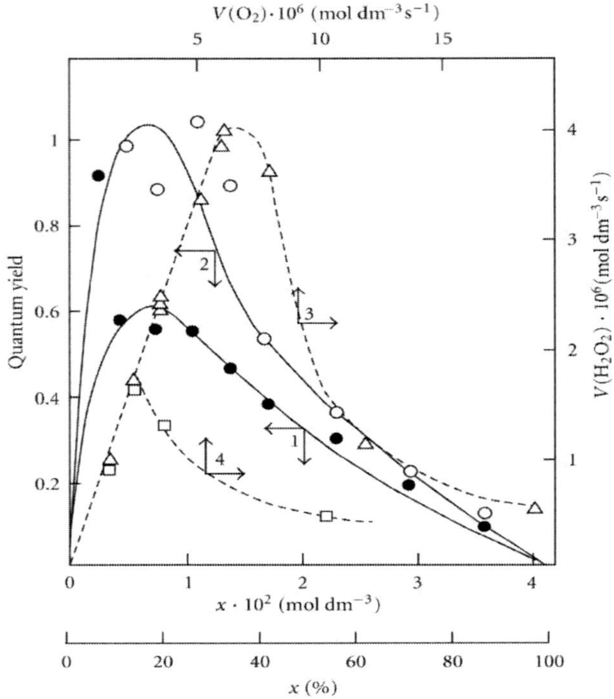

Figure 2. (*1*, *2*) Quantum yields of (*1*, ●) hydrogen peroxide and (*2*, ○) water resulting from the photochemical oxidation of hydrogen in the hydrogen–oxygen system as a function of the oxygen concentration *x* (light wavelength of 171.9–172.5 nm, total pressure of 10^5 Pa, room temperature [35]). (3, 4) Hydrogen peroxide formation rate $V(H_2O_2)$ (dashed curves) as a function of the rate $V(O_2)$ at which molecular oxygen is passed through a gas-discharge tube filled with (*3*, △) atomic and (*4*, □) molecular hydrogen. Atomic hydrogen was obtained from molecular hydrogen in the gas-discharge tube before the measurements (total pressure of 25–77 Pa, temperature of 77 K [12]). The symbols represent experimental data.

Curve *1* in Figure 1 illustrates the $RO_2^{\bullet}$ fit between Equation (1a) at $al >> \beta$ and experimental data for the radiation-induced oxidation of *o*-xylene in the liquid phase at 373 K in the case of 2-methylbenzyl hydroperoxide forming much more rapidly than *o*-tolualdehyde ($V_3 >> V_{3a}$ and $al >> \beta$, i.e., the parameter β can be neglected) [11]. The oxygen concentration limit in *o*-xylene is reached at an oxygen concentration of $[O_2]_{top} > x_m$, which corresponds to the third experimental point [11]. The oxygen concentration was calculated from the oxygen solubility in liquid xylene at 373 K [32]. The following quantities were used in this mathematical description: ^{60}Co γ-radiation dose rate of P = 2.18 Gy s^{-1} and total initiation yield of G(*o*-$CH_3C_6H_4\dot{C}H_2$) = 2.6 particles per 100 eV of the energy absorbed by the solution [11]; $V_1 = 4.73 \times 10^{-7}$ mol dm^{-3}s^{-1}, and $2k_5 = 1.15 \times 10^{10}$ dm^3 mol^{-1} s^{-1}. The resulting value of the parameter α is $(9.0 \pm 1.8) \times 10^{-3}$; hence, $k_2 = (3.2 \pm 0.8) \times 10^5$ dm^3 mol^{-1} s^{-1}. From the data presented in [33], it was estimated that $k_4 = k_3/\alpha = (5.2 \pm 1.2) \times 10^2$ dm^3 mol^{-1} s^{-1}.

Addition of the Hydrogen Atom

A number of experimental findings concerning the autoinhibiting effect of an increasing oxygen concentration at modest temperatures on hydrogen oxidation both in the gas phase [12, 35, 36] (Figure 2) and in the liquid phase [34] (Figure 1, curve *2*), considered in our earlier works [13, 27, 28, 37], can also be explained in terms of the competition kinetics of free radical addition [14, 38]. Figure 2 shows that the quantum yields of hydrogen peroxide and water (of products of photochemical oxidation of hydrogen at atmospheric pressure and room temperature) are maximum in the region of small concentrations of oxygen in the hydrogen–oxygen system (curves *1* and *2*, respectively) [35].

The hydroperoxyl free radical $HO_2^{\bullet}$ [46–50] resulting from reaction 2 possesses an increased energy due to the energy released in the conversion of the O=O multiple bond into the HO–O$^{\bullet}$ ordinary bond. Therefore, before its possible decomposition, it can interact with a hydrogen or oxygen molecule as the third body *via* parallel (competing) reactions 3 and 4, respectively. The hydroxyl radical HO$^{\bullet}$ that appears and disappears in consecutive parallel reactions 3 (first variant) and 3′ possesses additional energy owing to the exothermicity of the first variant of reaction 3, whose heat is distributed between the two products. As a consequence, this radical has a sufficiently high reactivity not to accumulate in the system during these reactions, whose rates are equal ($V_3 = V_{3'}$) under quasi-steady-state conditions, according to the above scheme. Parallel reactions 3 (second, parenthesized variant) and 3′ regenerate hydrogen atoms. It is assumed [27, 28] that the hydrotetraoxyl radical $HO_4^{\bullet}$ (first reported in [51–53]) resulting from endothermic reaction 4, which is responsible for the peak in the experimental rate curve (Figure 2, curve *3*), is closed into a five-membered $[\overset{\frown}{\mathrm{OO{-}H\cdots OO}}]^{\bullet}$ cycle due to weak intramolecular hydrogen bonding [25, 54]. This structure imparts additional stability to this radical and makes it least reactive.

Scheme 2

Nonbranched-chain oxidation of hydrogen and changes in enthalpy (ΔH°_{298}, kJ mol^{-1}) for elementary reactions[8].

Chain initiation:

$$1.\ H_2 \xrightarrow{h\nu,\ \gamma} 2H^{\bullet}, \qquad \Delta H^{o}_{298} = 436.0 \pm 0.0.$$

Chain propagation:

$$2.\ H^{\bullet} + O_2 \xrightarrow{k_2} HO_2^{\bullet}, \qquad \Delta H^{o}_{298} = -205.4 \pm 1.7;$$

$$3.\ HO_2^{\bullet} + H_2 \xrightarrow{k_3} H_2O + HO^{\bullet}, \qquad \Delta H^{o}_{298} = -215.4 \pm 2.9$$

$$(\text{or } H_2O_2 + H^{\bullet}), \qquad \Delta H^{o}_{298} = 69.4 \pm 1.7;$$

$$3'.\ HO^{\bullet} + H_2 \xrightarrow{k_{3'}} H_2O + H^{\bullet}, \qquad \Delta H^{o}_{298} = -62.8 \pm 1.2.$$

Inhibition:

$$4.\ HO_2^{\bullet} + O_2 \xrightarrow{k_4} HO_4^{\bullet}, \qquad \Delta H^{o}_{298} = 110.0 \pm 15.4.$$

Chain termination:

$$5.\ 2H^{\bullet}\ (+\ M) \xrightarrow{2k_5} H_2\ (+\ M), \qquad \Delta H^{o}_{298} = -436.0 \pm 0.0;$$

[8] According to Francisco and Williams [20], the enthalpy of formation ($\Delta H_f{}^{\circ}{}_{298}$) in the gas phase of $H^{\bullet}$, $HO^{\bullet}$, $HO_2^{\bullet}$, $HO_4^{\bullet}$ (the latter without the possible intramolecular hydrogen bond taken into account), O_3, H_2O [6], H_2O_2, and H_2O_4 is 218.0 ± 0.0, 39.0 ± 1.2, 12.6 ± 1.7, 122.6 ± 13.7, 143.1 ± 1.7, –241.8 ± 0.0, –136.0 ± 0, and –26.0 ± 9 kJ mol^{-1}, respectively. Calculations for the $HO_4^{\bullet}$ radical with a helical structure were carried out using the G2(MP2) method [39]. The stabilization energies of $HO_2^{\bullet}$, $HO_4^{\bullet}$, and $HO_3^{\bullet}$ were calculated in the same work to be 64.5 ± 0.1, 69.5 ± 0.8, and 88.5 ± 0.8 kJ mol^{-1}, respectively. The types of the O_4 molecular dimers, their IR spectra, and higher oxygen oligomers were reported [40, 41]. The structure and IR spectrum of the hypothetical cyclotetraoxygen molecule O_4, a species with a high energy density, were calculated by the CCSD method, and its enthalpy of formation was estimated [42]. The photochemical properties of O_4 and the van der Waals nature of the O_2–O_2 bond were investigated [43, 44]. The most stable geometry of the dimer is two O_2 molecules parallel to one another. The O_4 molecule was identified by NR mass spectrometry [45].

$$6.\ H^{\bullet} + HO_4^{\bullet} \xrightarrow{k_6} H_2O_2 + O_2, \qquad \Delta H^{o}_{298} = -476.6 \pm 13.7$$

$$(\text{or: } H_2O + O_3, \qquad \Delta H^{o}_{298} = -439.3 \pm 15.4,$$

$$H_2 + 2O_2), \qquad \Delta H^{o}_{298} = -340.6 \pm 13.7;$$

$$7.\ 2\,HO_4^{\bullet} \xrightarrow{2k_7} H_2O_2 + 2O_3, \qquad \Delta H^{o}_{298} = -95.0 \pm 30.8.$$

The $HO_4^{\bullet}$ radical was discovered by Staehelin *et al.* [55] in a pulsed radiolysis study of ozone degradation in water; its UV spectrum with an absorption maximum at 260 nm ($\varepsilon(HO_4^{\bullet})_{280\,nm}$= 320 ±15 m^2 mol^{-1}) was reported. The spectrum of the $HO_4^{\bullet}$ radical is similar to that of ozone, but the molar absorption coefficient $\varepsilon(HO_4^{\bullet})_{\lambda_{max}}$ of the former is almost two times larger [55]. The assumption about the cyclic structure of the $HO_4^{\bullet}$ radical can stem from the fact that its mean lifetime in water at 294 K, which is (3.6 ± 0.4) × 10^{-5} s (as estimated [37] from the value of 1/*k* for the monomolecular decay reaction $HO_4^{\bullet} \xrightarrow{k} HO_2^{\bullet} + O_2$ [55]), is 3.9 times longer than that of the linear $HO_3^{\bullet}$ radical [39, 56] estimated in the same way [37] for the same conditions [57], (9.1 ± 0.9) × 10^{-6} s.

MP2/6-311++G** calculations using the Gaussian-98 program confirmed that the cyclic structure of $HO_4^{\bullet}$ [58] is energetically more favorable than the helical structure [39] (the difference in energy is 4.8–7.3 kJ mol^{-1}, depending on the computational method and the basis set).[9] For example, with the MP2(full)/6-31G(d) method, the difference between the full energies of the cyclic and acyclic $HO_4^{\bullet}$ conformers with their zero-point energies (ZPE) values taken into account (which reduces the energy difference by 1.1 kJ mol^{-1}) is –5.1 kJ mol^{-1} and the entropy of the acyclic-to-cyclic $HO_4^{\bullet}$ transition is ΔS°_{298}= –1.6 kJ mol^{-1} K^{-1}. Therefore, under standard conditions, $HO_4^{\bullet}$ can exist in both forms, but the cyclic structure is obviously dominant (87%, K_{eq} = 6.5) [58].

Reaction 4 and, to a much lesser degree, reaction 6 inhibit the chain process because they lead to inefficient consumption of its main participants, $HO_2^{\bullet}$ and $H^{\bullet}$.

---

[9] There were calculations for the two conformers (cis and trans) of the $HO_4^{\bullet}$ radical [59] using large scale ab initio methods and density functional techniques with extended basis sets. Both conformers have a nearly planar geometry with respect to the four oxygen atoms and present an unusually long central O–O bond. The most stable conformer of $HO_4^{\bullet}$ radical is the cis one, which is computed to be endothermic with respect to $HO_2^{\bullet}$ (X^2A'') + O_2($X^3\Sigma_g^-$) at 0 K.

The hydrogen molecule that results from reaction 5 in the gas bulk possesses an excess energy, and, to acquire stability within the approximation used in this work, it should have time for deactivation *via* collision with a particle M capable of accepting the excess energy [60]. To simplify the form of the kinetic equations, it was assumed that the rate of the bimolecular deactivation of the molecule substantially exceeds the rate of its monomolecular decomposition, which is the reverse of reaction 5 [6].

Reactions 6 and 7 (taking into account the principle of detailed balance for the various pathways) regenerate hydrogen and oxygen (in the form of $O_2(X^3\Sigma_g^-)$ molecules, including the singlet states with: ΔH°_{f298} (O_2, $a^1\Delta_g$) = 94.3 kJ mol[-1] [20, 41] and ΔH°_{f298} (O_2, $b^1\Sigma_g^+$) = 161.4 kJ mol[-1] [41], which are deactivated by collisions, and in the form of O_3) and yield hydrogen peroxide or water *via* a nonchain mechanism, presumably through the intermediate formation of the unstable hydrogen tetraoxide molecule H_2O_4 [61, 62].[10] Ozone does not interact with molecular hydrogen. At moderate temperatures, it decomposes fairly slowly, particularly in the presence of $O_2(X^3\Sigma_g^-)$ [41]. The reaction of ozone with $H^{\bullet}$ atoms, which is not impossible, results in their replacement with $HO^{\bullet}$ radicals. The relative contributions from reactions 6 and 7 to the process kinetics can be roughly estimated from the corresponding enthalpy increments (Scheme 2).

When there is no excess hydrogen in the hydrogen–oxygen system and the homomolecular dimer O_4 [42–45, 63, 64], which exists at low concentrations (depending on the pressure and temperature) in equilibrium with O_2 [41], can directly capture the $H^{\bullet}$ atom to yield the heteronuclear cluster $HO_4^{\bullet}$,[11] which is more stable than O_4 [41] and cannot abstract a hydrogen atom from the hydrogen molecule, nonchain hydrogen oxidation will occur to give molecular oxidation products *via* the disproportionation of free radicals.

The low-reactive hydrotetraoxyl radical $HO_4^{\bullet}$ [55], which presumably has a high energy density [42], may be an intermediate in the efficient absorption and conversion of biologically hazardous UV radiation energy in the Earth's upper atmosphere. The potential energy surface for the atmospheric reaction

[10] The planar, six-atom, cyclic, hydrogen-bonded dimer $(HO_2^{\bullet})_2$ was calculated using quantum chemical methods (B3LYP density functional theory) [62]. The hydrogen bond energy is 47.7 and 49.4 kJ mol–1 at 298 K for the triplet and singlet states of the dimer, respectively.

[11] It is impossible to make a sharp distinction between the two-step bimolecular interaction of three species via the equilibrium formation of the labile intermediate O_4 and the elementary trimolecular reaction $O_2 + O_2 + H\bullet \rightarrow HO_4^{\bullet}$.

$HO^{\bullet} + O_3$, in which the adduct $HO_4^{\bullet}$ (^2A) was considered as an intermediate, was calculated by the double many-body expansion (DMBE) method [65]. From this standpoint, the following reactions are possible in the upper troposphere, as well as in the lower and middle stratosphere, where most of the ozone layer is situated (altitude of 16–30 km, temperature of 217–227 K, pressure of 1.0×10^4–1.2×10^3 Pa [66]; the corresponding ΔH°_{298} reaction values are given in kJ mol^{-1} [20]):

8. $H_2O(\text{vapor}) + h\nu \rightarrow H^{\bullet} + HO^{\bullet}$ [66]

9. $HO^{\bullet} + O_3 \rightarrow HO_4^{\bullet}$ [53, 55, 65], $\Delta H^{o}_{298} = -59.5$;

10. $HO_4^{\bullet} \rightarrow HO_2^{\bullet} + O_2(X^3\Sigma_g^-)$ [55, 65], $\Delta H^{o}_{298} = -110.0$

(or $HO_2^{\bullet} + O_2(a^1\Delta_g)$), $\Delta H^{o}_{298} = -15.7$.

The $HO_4^{\bullet}$ radical can disappear *via* disproportionation with a molecule, free radical, or atom in addition to dissociation. Note that emission from $O_2(a^1\Delta_g)$ and $O_2(b^1\Sigma_g^+)$ is observed at altitudes of 30–80 and 40–130 km, respectively [67].

Staehelin *et al.* [55] pointed out that, in natural systems in which the concentrations of intermediates are often very low, kinetic chains in chain reactions can be very long in the absence of scavengers since the rates of the chain termination reactions decrease with decreasing concentrations of the intermediates according to a quadratic law, whereas the rates of the chain propagation reactions decrease according to a linear law.

The kinetic description of the noncatalytic oxidation of hydrogen, including in an inert medium [60], in terms of the simplified scheme of free-radical nonbranched-chain reactions (Scheme 2), which considers only quadratic-law chain termination and ignores the surface effects [12], at moderate temperatures and pressures, in the absence of transitions to unsteady-state critical regimes, and at a substantial excess of the hydrogen concentration over the oxygen concentration was obtained by means of quasi-steady-state treatment, as in the previous studies on the kinetics of the branched-chain free-radical oxidation of hydrogen [47], even though the applicability of this method in the latter case under unsteady states conditions was insufficiently substantiated. The method was used with the following condition:[12] $k_6 = \sqrt{2k_5 2k_7}$ [1, 8].

The equation $V_3(H_2O_2; H_2O) = V_{3'}(H_2O)$ for the rate of the chain formation of hydrogen peroxide in the reaction (3) and water in reactions 3 and 3′ with $V_{3,3'}(H_2O) = 2V_3$ is identical to Equations (1, 1a) with the corresponding analytical expression for k_2 provided that $\beta = 0$ everywhere. The ratio of the rates of the competing reactions is $V_3/V_4 = \alpha l/x$, and the chain length is $\nu = V_3/V_1$. The equations for the rates of nonchain formation of hydrogen peroxide and water *via* reactions 6 and 7 – quadratic-law chain termination – are identical to Equations (4) and (5) assuming that $\beta = 0$. In these equations, l and x are the molar concentrations of hydrogen and oxygen ($l >> x$), l_m and x_m are the respective concentrations at the maximum point of the function, V_1 is the rate of initiation (reaction 1), $\alpha = k_3/k_4$, the rate constant:

$$k_2 = \alpha l_m \sqrt{2k_5 V_1} \Big/ x_m^2$$

is derived from the condition $\partial V_3/\partial x = 0$, and $2k_5$ is the rate constant of reaction 5 (hydrogen atom recombination), which is considered as bimolecular within the given approximation.

The rate constant $2k_5$ in the case of the pulsed radiolysis of ammonia–oxygen (+ argon) gaseous mixtures at a total pressure of 10^5 Pa and a temperature of 349 K was calculated to be 1.6×10^8 dm^3 mol^{-1} s^{-1} [36] (a similar value of this constant for the gas phase was reported in an earlier publication [68]). Pagsberg *et al.* [36] found that the dependence of the yield of the intermediate $HO^{\bullet}$ on the oxygen concentration has a maximum close to 5×10^{-4} mol dm^{-3}. In the computer simulation of the process, they considered the strongly exothermic reaction $HO_2^{\bullet} + NH_3 \rightarrow H_2O + {}^{\bullet}NHOH$, which is similar to reaction 3 in Scheme 2, whereas the competing reaction 4 was not taken into account.

The rate of the nonbranched-chain free-radical oxidation of hydrogen is a complex function of the rates of formation and disappearance of $H^{\bullet}$ atoms and $HO_4^{\bullet}$ radicals: $V_1 + V_{3,3'} - V_4 - V_5 + V_7$. Unlike the dependences of the rates

[12] For example, the ratio of the rate constants of the bimolecular disproportionation and dimerization of free radicals at room temperature is $k(HO^{\bullet} + HO_2^{\bullet})/[2k(2HO^{\bullet}) \times 2k(2HO_2^{\bullet})]^{0.5} = 2.8$ in the atmosphere [66] and $k(H^{\bullet} + HO^{\bullet})/[2k(2H^{\bullet})2k(2HO^{\bullet})]^{0.5} = 1.5$ in water [4]. These values that are fairly close to unity.

V_4 ($V_4 \leq V_1$), V_5 and V_7, the dependences of the rates V_2, $V_{3,\,3'}$, and $2V_6$ on the oxygen concentration x show a maximum.

Equation (1) with $\beta = 0$ under the conditions $k_2x^2 >> (\alpha l + x)\sqrt{2k_5V_1}$ and $k_2x^2 << (\alpha l + x)\sqrt{2k_5V_1}$, $\alpha l >> x$, corresponding to the descending and ascending branches of the curve with a maximum, can be transformed into simple equations which allow preliminary to estimate the parameters α and k_2 and express inversely (see Equation (6)) and directly proportional functions of concentration x:

$$V_3 = \sqrt{V_1}k_2x\big/\sqrt{2k_5}\,. \qquad (8)$$

In the case of nonchain hydrogen oxidation *via* the above addition reaction ($H^{\bullet} + O_4 \xrightarrow{k_{add}} HO_4^{\bullet}$), the formation rates of the molecular oxidation products in reactions 6 and 7 (Scheme 2, $k_2 = k_3 = k_4 = 0$) are defined by modified Equations (4) and (5) in which $\beta = 0$, $(\alpha l + x)$ is replaced with 1, and k_2 is replaced with $k_{add}K_{eq}$ ($k_{add}K_{eq}$ is the effective rate constant of $H^{\bullet}$ addition to the O_4 dimer, $K_{eq} = k/k'$ is the equilibrium constant of the reversible reaction $2O_2 \underset{k'}{\overset{k}{\Leftrightarrow}} O_4$ with $k' >> k_{add}[H^{\bullet}]$). The formation rates of the stable products of nonchain oxidation ($k_3 = 0$), provided that either reactions 2 and 4 or reaction 2 alone ($k_4 = 0$) occurs (Scheme 2; in the latter case, reactions 6 and 7 involve the $HO_2^{\bullet}$ radical rather than $HO_4^{\bullet}$), are given by modified Equations (4) and (5) with $\beta = 0$, $(\alpha l + x)$ replaced with 1, and x^2 replaced with x.

Note that, if in Scheme 2 chain initiation *via* reaction 1 is due to the interaction between molecular hydrogen and molecular oxygen yielding the hydroxyl radical $HO^{\bullet}$ instead of $H^{\bullet}$ atoms and if this radical reacts with an oxygen molecule (reaction 4) to form the hydrotrioxyl radical $HO_3^{\bullet}$ (which was obtained in the gas phase by neutralization reionization (NR) mass spectrometry [56] and has a lifetime of $>10^{-6}$ s at 298 K) and chain termination takes place *via* reactions 5–7 involving the $HO^{\bullet}$ and $HO_3^{\bullet}$, radicals instead of $H^{\bullet}$ and $HO_4^{\bullet}$, respectively, the expressions for the water chain formation rates derived in the same way will appear as a rational function of the oxygen concentration x without a maximum: $V_{3'}(H_2O) = V_1k_{3'}l\big/(k_4x + \sqrt{2k_5V_1})$.

Curve 2 in Figure 1 describes, in terms of the overall equation $V_{3,\,7} = V_1x(\alpha lf_m + x^3)\big/f_m^2$ for the rates of reactions 3 and 7 (which was derived from Equations (1a) and (5), respectively, with $\beta = 0$, Equation (5) in the form

[69] of $V_7 = V_1 x^4 / f_m^2$ (5a) in which k_2 is replaced with its analytical expression $\alpha l_m \sqrt{2k_5 V_1} / x_m^2$ derived from Equation (1) with β = 0 everywhere), the dependence of the hydrogen peroxide formation rate (minus the rate $V_{H_2O_2}$= 5.19×10^{-8} mol dm^{-3} s^{-1} of the primary formation of hydrogen peroxide after completion of the reactions in spurs) on the concentration of dissolved oxygen during the γ-radiolysis of water saturated with hydrogen (at the initial concentration 7×10^{-4} mol dm^{-3}) at 296 K [34].

This data was calculated in the present work from the initial slopes of hydrogen peroxide buildup versus dose curves for a ^{60}Co γ-radiation dose rate of P = 0.67 Gy s^{-1} and absorbed doses of $D \cong$ 22.5–304.0 Gy. The following values of the primary radiation-chemical yield G (species per 100 eV of energy absorbed) for water γ-radiolysis products in the bulk of solution at pH 4–9 and room temperature were used (taking into account that $V = GP$ and $V_1 = G_H P$): $G_{H_2O_2}$= 0.75 and G_H = 0.6 (initiation yield) [4]; $V_1 = 4.15 \times 10^{-8}$ mol dm^{-3} s^{-1}; $2k_5 = 2.0 \times 10^{10}$ dm^3 mol^{-1} s^{-1} [4]. As can be seen from Figure 1, the best description of the data with an increase in the oxygen concentration in water is attained when the rate V_7 of the formation of hydrogen peroxide *via* the nonchain mechanism in the chain termination reaction 7 (curve *1*, $\alpha = (8.5 \pm 2) \times 10^{-2}$) is taken into account in addition to the rate V_3 of the chain formation of this product *via* the propagation reaction 3 (dashed curve *2*, $\alpha = 0.11 \pm 0.026$). The rate constant of addition reaction 2 determined from α is substantially underestimated: $k_2 = 1.34 \times 10^7$ (*vs* 2.0×10^{10} [4]) dm^3 mol^{-1} s^{-1}. The difference can be due to the fact that the radiation-chemical specifics of the process were not considered in the kinetic description of the experimental data. These include oxygen consumption *via* reactions that are not involved in the hydrogen oxidation Scheme 2 and reverse reactions resulting in the decomposition of hydrogen peroxide by intermediate products of water radiolysis (e^-_{aq}, H$^\bullet$, HO$^\bullet$), with the major role played by the hydrated electron [4].

Conclusion

The above data concerning the competition kinetics of the nonbranched-chain addition of hydrocarbon free radicals and hydrogen atoms to the multiple bonds of the oxygen molecules make it possible to describe, using rate Equations (1a) and (5a), obtained by quasi-steady-state treatment, the peaking experimental dependences of the formation rates of molecular 1:1 adducts (in this case the o-$CH_3C_6H_4CH_2O_2H$ or H_2O_2) on the concentration of the oxygen over the entire range of its variation in binary systems (Figure 1). In such reaction systems consisting of saturated and unsaturated components [1, 13–15, 27, 28, 37, 38, 69–72], the unsaturated compound (in this case the O_2) is both a reactant and an autoinhibitor, specifically, a source of low-reactive free radicals (in this case the $RO_4^{\bullet}$ or $HO_4^{\bullet}$ radicals) shortening kinetic chains.

The progressive inhibition of the nonbranched-chain processes, which takes place as the concentration of the unsaturated compound is raised (after the maximum process rate is reached), can be an element of the self-regulation of the natural processes that returns them to the stable steady state.

Using the mechanism of the nonbranched-chain free-radical hydrogen oxidation considered here, it has been demonstrated that, in the Earth's upper atmosphere, the decomposition of O_3 in its reaction with the $HO^{\bullet}$ radical can occur *via* the addition of the latter to the ozone molecule, yielding the $HO_4^{\bullet}$ radical, which is capable of efficiently absorbing UV radiation [55].

References

[1] Silaev, M. M., Bugaenko, L. T. Mathematical Simulation of the Kinetics of Radiation Induced Hydroxyalkylation of Aliphatic Saturated Alcohols. *Radiation Physics and Chemistry*. 1992, *40*(1), 1–10.

[2] Vereshchinskii, I. V., Pikaev, A. K. *Vvedenie v radiatsionnuyu khimiyu* (*Introduction to Radiation Chemistry*); Spitsyn, V.I., Ed.; Akademiya Nauk SSSR: Moscow, 1963, p. 190.

[3] Silaev, M. M. Applied Aspects of the γ-Radiolysis of C_1–C_4 Alcohols and Binary Mixtures on Their Basis. *Khimiya Vysokikh Energii*. 2002, *36*(2), 97–101 (English Translation in: *High Energy Chemistry*. 2002, *36*(2), 70–74.

[4] Pikaev, A. K. *Sovremennaya radiatsionnaya khimiya. Radioliz gazov i zhidkostei* [*Modern Radiation Chemistry: Radiolysis of Gases and Liquids*]; Nauka: Moscow, 1986, 440 pp.

[5] Emanuel, N. M., Denisov, E. T., Maizus, Z. K. *Tsepnye reaktsii okisleniya uglevodorodov v zhidkoi faze* [*Chain Oxidation Reactions of Hydrocarbons in the Liquid Phase*], Nauka: Moscow, 1965, 375 pp.

[6] Benson, S. W. *Thermochemical Kinetics: Methods for the Estimation of Thermochemical Data and Rate Parameters*; 2nd Edition, Wiley: New York, 1976, 320 pp.

[7] Walling, Ch. *Free Radicals in Solution*; Wiley: New York, 1956, 631 pp.

[8] Bateman, L. Olefin Oxidation. *Quarterly Reviews*. 1954, *8*(2), 147–167.

[9] Shtern, V. Ya. *Mekhanizm okisleniya uglevodorodov v gazovoi faze* [*Mechanism of the Gas-Phase Oxidation of Hydrocarbons*]; Akademiya Nauk SSSR: Moscow, 1960, 496 pp.

[10] Bäckström, H. L. J. Der Kettenmechanismus bei der Autoxydation von Aldehyden [*The chain mechanism in the autoxidation of aldehydes*]. *Zeitschrift für physikalische Chemie (B).* 1934, *25*(1–2), 99–121.

[11] Aliev, A. A., Saraeva, V. V. Isomerization of Peroxy Radicals Resulting from the Radiation-Induced Oxidation of *o*-Xylene. *Vestnik Moskovskogo Universiteta*; Ser. 2: Khimiya. 1983, *34*(4), 371–374.

[12] Badin, E. J. The Reaction between Atomic Hydrogen and Molecular Oxygen at Low Pressures. Surface Effects. *Journal of the American Chemistry Society.* 1948, *70*(11), 3651–3655.

[13] Silaev, M. M. A New Competitive Kinetic Model of Radical Chain Oxidation: Oxygen as an Autoinhibitor. *Biofizika.* 2001, *46*(2), 203–209 (English Translation in: *Biophysics*. 2001, *46*(2), 202–207.

[14] Silaev, M. M. Simulation of the Initiated Addition of Hydrocarbon Free Radicals and Hydrogen Atoms to Oxygen *via* a Nonbranched Chain Mechanism. *Teoreticheskie Osnovy Khimicheskoi Tekhnologii*. 2007, *41*(6), 634–642. (English Translation in: *Theoretical Foundation of Chemical Engineering*). 2007, *41*(6), 831–838.

[15] Silaev, M. M., Bugaenko, L. T. Kinetics of the Addition of α-Hydroxyalkyl Radicals to 2-Propen-1-ol and Formaldehyde. *Kinetics and Katalysis.* 1994, *35*(4), 509–513.

[16] Gurvich, L. V., Karachevtsev, G. V., Kondrat'ev, V. N., Lebedev, Yu. A., Medvedev, V. A., Potapov, V. K., Khodeev, Yu. S. *Energii razryva khimicheskikh svyazei. Potentsialy ionizatsii i srodstvo k elektronu* [*Bond Dissociation Energies, Ionization Potentials, and Electron Affinity*]; Kondrat'ev, V. N.; Ed.; Nauka: Moscow, 1974, 351 pp.

[17] Orlov, Yu. D., Lebedev, Yu. A., Saifullin, I. Sh. *Termokhimiya organicheskikh svobodnykh radikalov* [*Thermochemistry of Organic Free Radicals*]; Kutepov, A. M., Ed.; Nauka: Moscow, 2001, 304 pp.

[18] Pedley, J. B., Naylor, R. D., Kirby, S. P. *Thermochemical Data of Organic Compounds*; 2nd Edition; Chapman & Hall: London, 1986, 792 pp.

[19] Buchachenko, A. L. *Kompleksy radikalov i molekulyarnogo kisloroda s organicheskimi molekulami* [*Complexes of Radicals and Dioxygen with Organic Molecules*]; Beletskaya, I. P.; Ed.; Nauka: Moscow, 1984, 158 pp.

[20] Francisco, J. S., Williams, I. H. The Thermochemistry of Polyoxides and Polyoxy Radicals. *International Journal of Chemical Kinetics.* 1988, *20*(6), 455–466.

[21] Kokorev, V. N., Vyshinskii, N. N., Maslennikov, V. P., Abronin, I. A., Zhidomirov, G. M., Aleksandrov, Yu. A. Electronic Structure and Chemical Reactions of Peroxides: I. MINDO/3 Calculation of the Geometry and Enthalpy of Formation of the Ground States of Organic and Organoelement Peroxides. *Zhurnal Strukturnoi Khimii*. 1981, *22*(4), 9–15.

[22] Dmitruk, A. F., Lobanov, V. V., Kholoimova, L. I. Role of Tetroxide Conformation in the Mechanism of Peroxy Radical Recombination. *Teoreticheskaya i Eksperimental'naya Khimiya*. 1986, *22*(3), 363–366.

[23] Belyakov, V. A., Vasil'ev, R. F., Ivanova, N. M., Minaev, B. F., Osyaeva, O. V., Fedorova, G. F. Electronic Model of the Excitation of Chemiluminescence in the Oxidation of Organic Compounds. *Izvestiya Akademii Nauk SSSR; Ser.: Fizika*. 1987, *51*(3), 540–547.

[24] Ase, P., Bock, W., Snelson, A. Alkylperoxy and Alkyl Radicals. 1. Infrared Spectra of CH_3O_2 and $CH_3O_4CH_3$ and the Ultraviolet Photolysis of CH_3O_2 in Argon + Oxygen Matrices. *The Journal of Physical Chemistry*. 1986, *90*(10), 2099–2109.

[25] Pimentel, G. C., McClellan, A. L. *The Hydrogen Bond*; Pauling, L., Ed.; Freeman: San Francisco, 1960, p. 200.

[26] Russell, G. A. Deuterium-Isotope Effects in the Autooxidation of Aralkyl Hydrocarbons: Mechanism of the Interaction of Peroxy Radicals. *Journal of the American Chemical Society*. 1957, *79*(14), 3871–3877.

[27] Silaev, M. M. The Competition Kinetics of Nonbranched Chain Processes of Free-Radical Addition to Double Bonds of Molecules with the Formation of 1:1 Adducts and the Inhibition by the Substrate. *Oxidation Communication*. 1999, *22*(2), 159–170.

[28] Silaev, M. M. The Competition Kinetics of Radical-Chain Addition. *Zhurnal Fizicheskoi Khimii*. 1999, *73*(7), 1180–1184 (English Translation in: *Russian Journal of Physical Chemistry*. 1999, *73*(7), 1050–1054.

[29] Darmanyan, A. P., Gregory, D. D., Guo, Y., Jenks, W. S., Burel, L., Eloy, D., Jardon, P. Quenching of Singlet Oxygen by Oxygen- and Sulfur-Centered Radicals: Evidence for Energy Transfer to Peroxy Radicals in Solution. *Journal of the American Chemistry Society*. 1998, *120*(2), 396–403.

[30] Kanofsky, J. R. Singlet Oxygen Production from the Reactions of Alkylperoxy Radicals. Evidence from 1268-nm Chemiluminescence. *The Journal of Organic Chemistry*. 1986, *51*(17), 3386–3388.

[31] Semenov, N. N. *Tsepnye reaktsii* (*Chain Reactions*); Goskhimtekhizdat, Leningrad, 1934, pp. 241, 203.

[32] Reznikovskii, M., Tarasova, Z., Dogadkin, B. Oxygen Solubility in Some Organic Liquids. *Zhurnal Obshchei Khimii*. 1950, *20*(1), 63–67.

[33] Howard, J. A., Ingold, K. U. Absolute Rate Constants for Hydrocarbon Autooxidation. VI. Alkyl Aromatic and Olefinic Hydrocarbons. *Canadian Journal of Chemistry*. 1967, *45*(8), 793–802.

[34] Barr, N. F., Allen, A. O. Hydrogen Atoms in the Radiolysis of Water. *The Journal of Physical Chemistry*. 1959, *63*(6), 928–931.

[35] Smith, H. A., Napravnik, A. Photochemical Oxidation of Hydrogen. *Journal of the American Chemistry Society*. 1940, *62*(1), 385–393.

[36] Pagsberg, P. B., Eriksen, J., Christensen, H. C. Pulse Radiolysis of Gaseous Ammonia–Oxygen Mixtures. *The Journal of Physical Chemistry*. 1979, *83*(5), 582–590.

[37] Silaev, M. M. Competitive Mechanism of the Non-branched Radical Chain Oxidation of Hydrogen Involvingthe Free Cyclohydrotetraoxyl Radical

[$\overline{OO\cdots H\cdots OO}$]$^{\bullet}$, which Inhibits the Chain Process. *Khimiya Vysokikh Energii*. 2003, 37(1), 27–32 (English Translation in: *High Energy Chemistry*. 2003, *37*(1), 24–28.

[38] Silaev, M. M. Simulation of Initiated Nonbranched Chain Oxidation of Hydrogen: Oxygen as an Autoinhibitor. *Khimiya Vysokich Energii*. 2008, *42*(2), 124–129 (English Translation in: *High Energy Chemistry*. 2008, *42*(*1*), 95–100.

[39] McKay, D. J., Wright, J. S. How Long Can You Make an Oxygen Chain? *Journal of the American Chemistry Society*. 1998, *120*(5), 1003–1013.

[40] Lipikhin, N. P. Dimers, Clusters, and Cluster Ions of Oxygen in the Gas Phase. *Uspekhi Khimii*. 1975, *44*(8), 366–376.

[41] Razumovskii, S. D. *Kislorod – elementarnye formy i svoistva* [*Oxygen: Elementary Forms and Properties*]; Khimiya: Moscow, 1979, 302 pp.

[42] Dunn, K. M., Scuceria, G. E., Schaefer III, H. F. The infrared spectrum of cyclotetraoxygen, O_4: a theoretical investigation employing the single and double excitation coupled cluster method. *The Journal of Chemical Physics*. 1990, *92*(10), 6077–6080.

[43] Brown, L., Vaida, V. Photoreactivity of Oxygen Dimers in the Ultraviolet. *The Journal of Physical Chemistry*. 1996, *100*(19), 7849–7853.

[44] Aquilanti, V., Ascenzi, D., Bartolomei, M., Cappelletti, D., Cavalli, S., de Castro-Vitores, M., Pirani, F. Molecular Beam Scattering of Aligned Oxygen Molecules. The Nature of the Bond in the O_2–O_2 Dimer. *Journal of the American Chemistry Society*. 1999, *121*(46), 10794–1080.

[45] Cacace, F., de Petris, G., Troiani, A. Experimental Detection of Tetraoxygen. *Angewandte Chemie*; *Internation Edition* (in English). 2001, *40*(21), 4062–4065.

[46] Taylor, H. S. Photosensitisation and the Mechanism of Chemical Reactions. *Transactions of the Faraday Society*. 1926, *21*(63/3), 560–568.

[47] Nalbandyan, A. B., Voevodskii, V. V. *Mekhanizm okisleniya i goreniya vodoroda* [*Mechanism of Hydrogen Oxidation and Combustion*]; Kondrat'ev, V. N.; Ed.; Akademiya Nauk SSSR: Moscow–Leningrad, 1949, 180 pp.

[48] Foner, S. N., Hudson, R. L. Mass spectrometry of the HO_2 free radical. *The Journal of Chemical Physics*. 1962, *36*(10), 2681.

[49] Hochanadel, C. J., Ghormley, J. A., Ogren, P. J. Absorption Spectrum and Reaction Kinetics of the HO_2 Radical in the Gas Phase. *The Journal of Chemical Physics*. 1972, *56*(9), 4426–4432.

[50] Lightfoot, P. D., Veyret, B., Lesclaux, R. Flash Photolysis Study of the CH_3O_2 + HO_2 Reaction between 248 and 573 K. *The Journal of Physical Chemistry*. 1990, *94*(2), 708–714.

[51] Smith, P. The HO_3 and HO_4 Free Radicals. *Chemistry and Industry*. 1954, (42), 1299–1300.

[52] Robertson, A. J. B. A Mass Spectral Search for H_2O_4 and HO_4 in a Gaseous Mixture Containing HO_2 and O_2. *Chemistry and Industry*. 1954, (48), 1485.

[53] Bahnemann, D., Hart, E. J. Rate Constants of the Reaction of the Hydrated Electron and Hydroxyl Radical with Ozone in Aqueous Solution. *The Journal of Physical Chemistry*. 1982, *86*(2), 252–255.

[54] *Vodorodnaya svyaz': Sbornik statei* [*The Hydrogen Bonding: Collection of Articles*]; Sokolov, N. D.; Ed.; Nauka: Moscow, 1981, 288 pp.
[55] Staehelin, J., Bühler, R. E., Hoigné, J. Ozone Decomposition in Water Studied by Pulse Radiolysis. 2. OH and HO_4 as Chain Intermediates. *The Journal of Physical Chemistry*. 1984, *88*(24), 5999–6004.
[56] Cacace, F., de Petris, G., Pepi, F., Troiani, A. Experimental Detection of Hydrogen Trioxide. *Science*. 1999, *285*(5424), 81–82.
[57] Bühler, R. F., Staehelin, J., Hoigné, J. Ozone Decomposition in Water Studied by Pulse Radiolysis. 1. HO_2/O_2^- and HO_3/O_3^- as Intermediates. *The Journal of Physical Chemistry*. 1984, *88*(12), 2560–2564.
[58] Trushkov, I. V., Silaev, M. M., Chuvylkin, N. D. Acyclic and Cyclic Forms of the Radicals $HO_4^\bullet$, $CH_3O_4^\bullet$, and $C_2H_5O_4^\bullet$: Ab Initio Quantum Chemical Calculations. *Izvestiya Akademii Nauk; Ser.: Khimiya*. 2009, (3), 479–482 (English Translation in: *Russian Chemical Bulletin; International Edition*. 2009, *58*(3), 489–492).
[59] Mansergas, A., Anglada, J. M., Olivella, S., Ruiz-López, M. F. On the Nature of the Unusually Long OO Bond in HO_3 and HO_4 Radicals. *Phys. Chem. Chem. Phys*. 2007, *9*(44), 5865–5873.
[60] Wong, W., Davis, D. D. A Flash Photolysis Resonance Fluorescence Study of the Reactions of Atomic Hydrogen and Molecular Oxygen: $H + O_2 + M \rightarrow HO_2 + M$. *International Journal of Chemical Kinetics*. 1974, *6*(3), 401–416.
[61] Yagodovskaya, T. V., Nekrasov, L. I. Use of Infrared Spectroscopy to Study Frozen Hydrogen–Oxygen Systems Containing Hydrogen Polyoxides (in Russian). *Zhurnal Fizicheskoi Khimii*. 1977, *51*(10), 2434-45.
[62] Xu, X., Muller, R. P., Goddard III, W. A. The Gas Phase Reaction of Singlet Dioxygen with Water: A Water-Catalyzed Mechanism. *Proceedings of the National Academy Sciences of the United States of America*. 2002, *99*(6), 3376–3381.
[63] Seidl, E. T., Schaefer III, H. F. Is There a Transition State for the Unimolecular Dissociation of Cyclotetraoxygen (O_4)? *The Journal of Chemical Physics*. 1992, *96*(2), 1176–1182.
[64] Hernández-Lamoneda, R., Ramírez-Solís, A. Reactivity and Electronic States of O_4 along Minimum Energy Paths. *The Journal of Chemical Physics*. 2000, *113*(10), 4139–4145.
[65] Varandas, A. J. C., Zhang, L. Test Studies on the Potential Energy Surface and Rate Constant for the $OH + O_3$ Atmospheric Reaction. *Chemical Physics Letters*. 2000, *331*(5–6), 474–482.
[66] *Atmosfera. Spravochnik* [*Atmosphere: A Handbook*]; Gidrometeoizdat: Leningrad, 1991, 510 pp.
[67] Okabe, H. *Photochemistry of Small Molecules;* Wiley, New York, 1978, 431 pp.
[68] Boyd, A. W., Willis, C., Miller, O. A. A Re-examination of the Yields in the High Dose Rate Radiolysis of Gaseous Ammonia. *Canadian Journal of Chemistry*. 1971, *49*(13), 2283–2289.
[69] Silaev, M. M. Competition Mechanism of Substrate-Inhibited Radical Chain Addition to Double Bond. *Neftekhimiya*, 2000, *40*(1), 33–40 (English Translation in: *Petroleum Chemistry*. 2000, *40*(1), 29–35.

[70] Silaev, M. M. Competition Kinetics of Nonbranched Chain Processes of Free Radical Addition to the C=C, C=O, and O=O Double Bonds of Molecules. *Neftekhimiya*. 2003, *43*(4), 302–307 (English Translation in: Petroleum Chemistry. 2003, *43*(4), 258–273.

[71] Silaev, M. M. Low-reactive Free Radicals Inhibiting Nonbranched Chain Processes of Addition. *Biofizika*. 2005, *50*(4), 585–600 (English Translation in: *Biophysics*. 2005, *50*(4), 511–524.

[72] Silaev, M. M. Kinetics of the Free-Radical Nonbranched-Chain Addition. *American Journal of Polymer Science and Technology*. 2017, *3*(3), 29-49.

Chapter 6

A Kinetic and Mechanistic Modeling on Oxidation of the Anti-Diabetic Drug Metformin with Cu(III) Periodate Complex in Aqueous Alkaline Media

Rajesh N. Hegde* **and P. Vishwanatha**
Department of Chemistry,
Sri Dharmasthala Manjunatheshwara College (Autonomous), Ujire, India

Abstract

The kinetics of oxidation of an antidiabetic drug, metformin by Copper (III) periodate complex in an aqueous alkaline medium at a constant ionic strength of 0.50 mol dm^{-3} was studied spectrophotometrically. The reaction between Copper (III) periodate complex and metformin in the KOH medium exhibits 1:6 stoichiometry (metformin: Copper (III) periodate complex). The reaction is of first order in [Copper (III)] and less than unit order dependence each in [MET] and [OH^-]. The rate of the reaction decreases with an increase in [periodate]. The rate of the reaction increases with an increase in the ionic strength of the medium. The effect of added products and the dielectric constant of the reaction medium were also examined. A suitable oxidation mechanism was proposed. The main product was identified by the LC-MS study. The reaction constants involved in the different steps of the mechanism are calculated. The activation parameters and thermodynamic quantities were computed by graphical method and discussed.

Keywords: oxidation, metformin, kinetics, Copper (III)

* Corresponding Author's Email: rajeshhegd@gmail.com.

In: A Closer Look at Chemical Kinetics
Editor: Victor Martinez-Luaces
ISBN: 979-8-88697-484-3

Introduction

The highest oxidation state exhibited by transition metals is of much interest among researchers, particularly in electron transfer reactions. Polydentate ligands are being used to stabilize the transition metal complexes with highest oxidation state via chelation. Various metal chelates such as diperiodatoargentate (III) [1], diperiodatocuprate (III) [2], and diperiodatonickelate (IV) [3] were used as good oxidizing agents in suitable pH ranges. Cu^{3+} complexes with tellurate and periodate have been used for the analysis of many organic compounds [4]. Due to the abundance and significance of copper complexes in biological chemistry, they have been used widely in oxidation reactions [5]. Owing to the solubility and stability of diperiodatocuprate (III) (DPC) in aqueous media, its use was restricted. DPC was recognized as a one-electron oxidant in aqueous alkaline medium [2]. The DPC has been used for the estimation of various metal ions such as chromium, magnesium and calcium present in the ores, and arsenic, antimony, and tin in their alloys [6]. It would be very much interesting to know which is the active species since Copper (III) periodate complex exists as multiple Copper (III) species in aqueous alkaline media [7].

The worldwide disease, Diabetes, has been classified as type 1 or insulin dependent and type 2 or non-insulin dependent. Metformin (N,N-dimethylbiguanide) (MET), is a biguanide prescribed and used to treat type 2 diabetes [8] and is known to increase glucose transport across the cell membrane in skeletal muscles. MET is the drug of choice for obese patients [9] and it can also inhibit the formation of advanced glycosylation end products [10]. According to International Diabetes Federation, one in ten people are living with diabetes. The statistical data says that approximately 463 million adults were living with diabetes worldwide and by 2045 this will rise to 700 million, the prediction says. People with diabetes are more vulnerable to the severe effect of coronavirus. MET has common adverse effects such as diarrhea, nausea, and abdominal pain and is not recommended for those with significant liver disease [11]. A survey of literature revealed that there was no study on the oxidation kinetics of MET with DPC. Because of the pharmaceutical importance of MET, we have undertaken the detailed kinetic and mechanistic study on the oxidation of this antidiabetic drug with soluble Copper (III) periodate complex.

Experiments

Reagent grade chemicals were used in the present study and doubly distilled water was used right through the study. The Copper (III) periodate complex was prepared by a known method [12] and standardized using a standard method [13]. The maximum absorption spectra at 415 nm in the UV-Visible spectra was used to confirm the existence of Copper (III) complex. The solutions of MET and copper sulfate were prepared by dissolving known quantities of sample in doubly distilled water. The periodate solution was prepared and standardized iodometrically [13]. The alkalinity and ionic strength were maintained in the reaction using KOH and KNO_3, respectively. In order to vary the dielectric constant of the medium, t-butyl alcohol was used.

For kinetic measurements, a Peltier Accessory (to control the temperature) attached Varian CARY 50 Bio UV-vis spectrophotometer (Varian, Victoria-3170, Australia) was used. For product analysis, LC-MS (Waters, Synapt G2) high detection mass spectrometer, using the Electron Ionization (EI) technique, was employed. For pH measurements, the Elico pH meter (model LI120) was used.

Additionally, using a Varian Carry 50 Bio UV-Visible spectrophotometer, kinetic measurements were done. Under pseudo-first order conditions, the kinetics was followed where [MET] > [DPC] at $27 \pm 0.1^{\circ}C$. By mixing MET and DPC solutions the reaction was initiated, containing required quantities of KOH, KNO_3 and KIO_4. By monitoring the decrease in absorbance value of DPC at 415 nm, the progress of the reaction was tracked. Prior to the kinetic runs, Beer's law was verified and the extinction coefficient was found to be $\varepsilon = 6240 \pm 100$ dm^3 mol^{-1} cm^{-1}. It was confirmed that the interference of other species present in the reaction mixture was negligible at this wavelength.

The rate constants, k_{obs} were obtained from the graph of $\log_{10}$(absorbance) against time and these constants were reproducible within ±5% error. The graphs were linear up to 75% completion of the reaction. To fix the constant concentration of DPC, a constant concentration was used throughout the kinetic runs. The possibility of oxidation of MET by periodate in aqueous alkaline media was also studied since the periodate concentration used was high. But it was confirmed that there was no noteworthy reaction between MET and periodate as compared to the oxidation of MET by DPC.

Results and Discussion

Different sets of reaction mixtures containing uneven ratios of DPC to MET in the existence of constant amounts of OH^- and KNO_3 were kept for 3 hours in a closed container under inert atmosphere. After 3 hours, the residual concentration of DPC was estimated spectrophotometrically at 415 nm. The results point out a 1:6 stoichiometry as given in Equation 1.

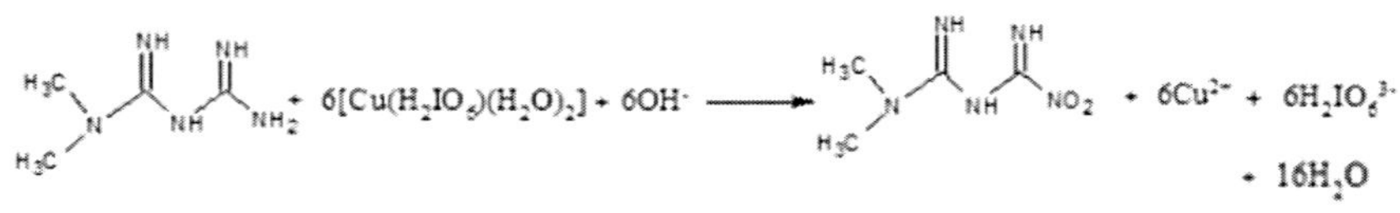

(1)

The reaction mixture of MET and DPC with the required amount of KOH solution was stirred for 36 hours at 27°C. The reaction products were neutralized with HCl and the aqueous solution present was extracted 4 times with ethyl acetate and was subjected to thin layer chromatographic analysis. The oxidation product of MET, N-(N,N-dimethyl carbamimidoyl)-1-nitro formimidamide, was confirmed by LC-MS analysis. LC-MS analysis of the reaction mixture indicated the presence of the product, N-(N,N-dimethyl carbamimidoyl)-1-nitro formimidamide, with molecular ion peak (M + 1) of *m/z* 160.13 amu (Figure 1). It was observed that N-(N,N-dimethyl carbamimidoyl)-1-nitro formimidamide, did not undergo further oxidation under the present kinetic conditions.

The concentrations of the oxidant, DPC were varied in the range of 1.0×10^{-5} to 1.0×10^{-4} mol dm^{-3} by keeping OH^-, MET and periodate concentrations constant. The order with respect to DPC concentration was unity as seen from the reasonably constant values of k_{obs} (Table 1). The plot of $\log_{10}$(absorbance) against time (Figure 2) was linear up to 75% completion of the reaction and it indicates that the order with respect to the oxidant was unity.

The influence of concentration of substrate, MET on the rate of reaction was studied by maintaining a constant concentration of DPC, alkali and IO_4^-. At a constant ionic strength of 0.50 mol dm^{-3}, the effect of the substrate was studied in the range of 1.0×10^{-4} to 1.0×10^{-3} mol dm^{-3}. The rate of the reaction increased with an increase in the concentration of MET and the order with respect to the substrate was found to be less than unity.

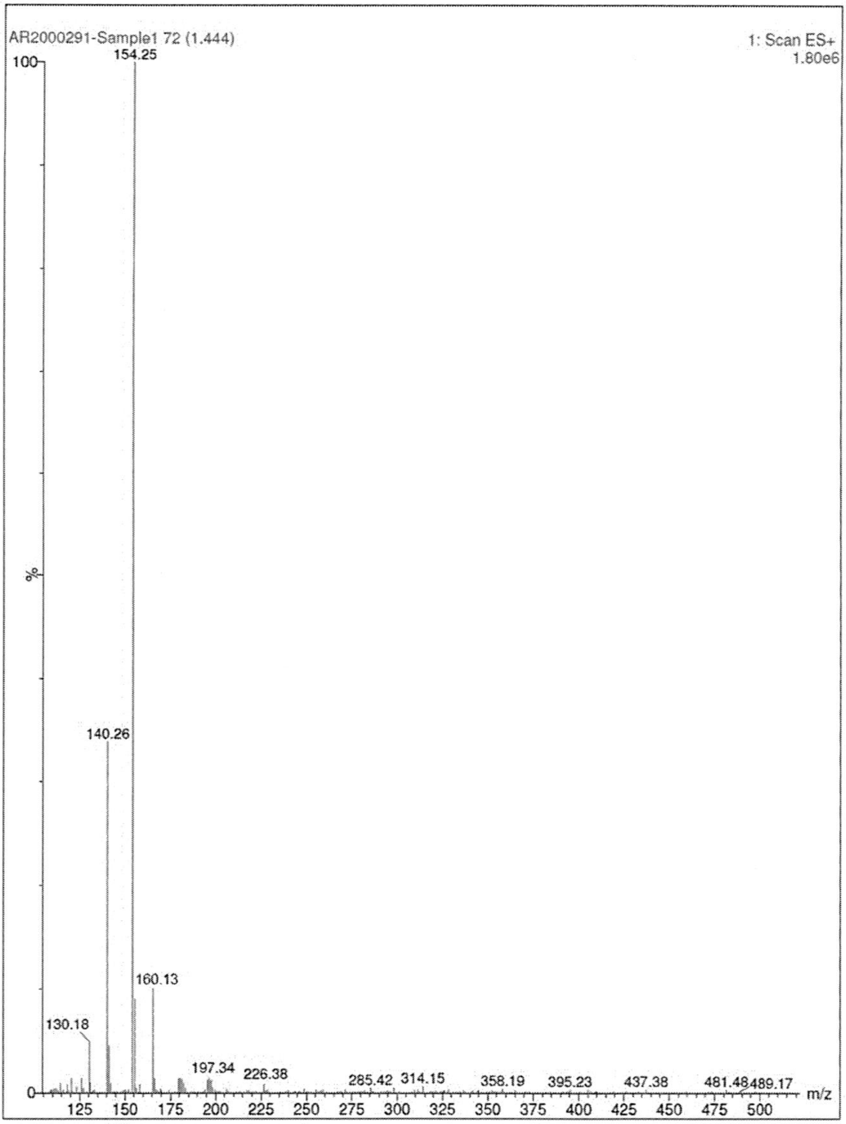

Figure 1. LC-MS spectrum of N-(N,N-dimethyl carbamimidoyl)-1-nitro formimidamide (M+1) m/z at 160.13 amu.

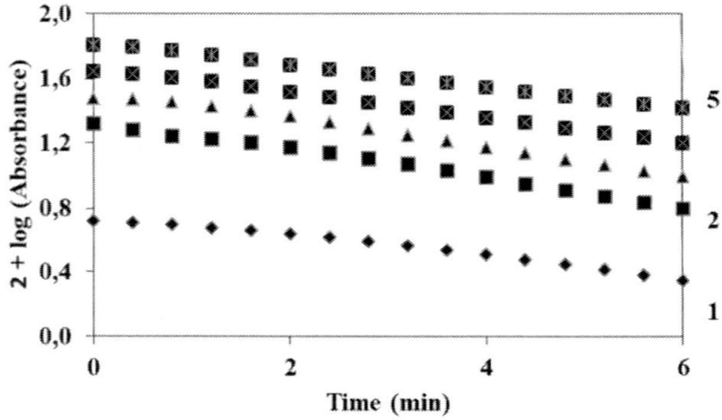

Figure 2. First order plots on the oxidation of MET with DPC in aqueous alkaline medium. [DPC]: (1) 1.0, (2) 3.0, (3) 5.0, (4) 7.0 and (5) 10.0 x 10^{-5} mol dm^{-3}.

Table 1. Influence of [DPC], [MET], [OH^-] and [IO_4^-] on the oxidation of MET with diperiodatocuprate (III) in aqueous alkaline medium at 27°C, I = 0.10 mol dm^{-3}

[DPC] x10^5 (mol dm^{-3})	[MET] x10^4 (mol dm^{-3})	[OH^-] (mol dm^{-3})	[IO_4^-] x10^5 (mol dm^{-3})	k_{obs} x10^3 (s^{-1}) Found	Calculated
1.0	5.0	0.01	5.0	2.21	2.13
3.0	5.0	0.01	5.0	2.32	2.13
5.0	5.0	0.01	5.0	2.13	2.13
7.0	5.0	0.01	5.0	2.33	2.13
10.0	5.0	0.01	5.0	2.03	2.13
5.0	1.0	0.01	5.0	0.53	0.56
5.0	2.0	0.01	5.0	1.11	1.04
5.0	5.0	0.01	5.0	2.13	2.13
5.0	8.0	0.01	5.0	2.97	2.89
5.0	10.0	0.01	5.0	3.16	3.27
5.0	5.0	0.006	5.0	1.73	1.76
5.0	5.0	0.008	5.0	1.91	1.97
5.0	5.0	0.01	5.0	2.13	2.13
5.0	5.0	0.02	5.0	2.44	2.53
5.0	5.0	0.06	5.0	2.92	2.86
5.0	5.0	0.01	1.0	3.60	3.56
5.0	5.0	0.01	2.0	3.12	3.05
5.0	5.0	0.01	5.0	2.13	2.13
5.0	5.0	0.01	7.0	1.75	1.77
5.0	5.0	0.01	10.0	1.36	1.42

The influence of the concentration of alkali was studied in the range of 6.0×10^{-3} to 6.0×10^{-2} mol dm^{-3} by keeping other concentrations constant at 27°C. The rate constant of the oxidation reaction was increased with increasing amounts of alkali (Table 1). The order with respect to OH^- concentration was positive and less than unity.

The concentration of periodate was varied from 1.0×10^{-5} to 1.0×10^{-4} mol dm^{-3} by maintaining all other concentrations and conditions constant. As seen from the k_{obs} values tabulated (Table 1), the rate of the reaction has decreased with an increase in the concentration of IO_4^-. The order with respect to the periodate concentration was found to be negative and fractional.

The effect of ionic strength on the reaction was studied by varying the quantity of potassium nitrate. The ionic strength of the reaction medium was varied from 0.1 to 0.5 mol dm^{-3} at constant concentrations of DPC, MET, periodate and alkali. It was observed that the k_{obs} values increased with increasing the concentration of KNO_3. A linear graph with a positive slope was obtained by plotting log k_{obs} versus $I^{1/2}$. Using different proportions of water and tert-butyl alcohol, the dielectric constant of the medium was varied. From the equation, $D = D_w V_w + D_B V_B$, the dielectric constant values 'D' were calculated [14]. The D_w and D_B are dielectric constants of pure water and tert-butyl alcohol respectively, and V_w and V_B are volume fractions of components of water and tert-butyl alcohol (varied between 5% to 40% of water – tert-butyl alcohol) respectively in the total mixture. The obtained k_{obs} values were fairly constant, indicating that the dielectric constant has no noteworthy influence on the rate of this oxidation reaction.

The rate of the reaction was not affected appreciably by the initially added product, Cu (II). From the above experimental observations, the rate law can be written as:

$$\text{Rate} = k_{obs}\,[\text{DPC}]\,[\text{MET}]^{0.72}\,[\text{OH}^-]^{0.53}\,[\text{IO}_4^-]^{-0.42}$$

The involvement of free radicals in the reaction was examined as follows. A known quantity of acrylonitrile monomer was added initially to the reaction mixture and was kept for 2 hours in an inert atmosphere. A white precipitate was obtained when the reaction mixture was diluted with methanol, indicating the involvement of free radicals in the reaction. It should be noted that the blank experiments of either DPC or MET alone with the monomer acrylonitrile did not induce any polymerization under the same condition as those induced for the reaction mixture. However, the rate of reaction decreased indicating a free radical involvement, when acrylonitrile was added initially, as reported [2].

In order to identify the effect of dissolved oxygen on the rate of reaction, the reaction was carried out in an inert atmosphere. No considerable divergence in the results was noticed under an inert atmosphere and in the presence of air. The effect of carbonate was examined on the rate of reaction, since carbonate contamination was common. However, the reaction rates were not affected by added carbonate.

The reaction orders were determined from the slopes of graphs of $\log_{10}k_{obs}$ against $\log_{10}$concentration by fluctuating the quantities of periodate, MET, and alkali, by keeping other conditions and quantities constant.

The Copper (III) periodate complex which is water soluble was reported to be $[Cu(HIO_6)_2]^{5-}$[15]. In the present study, the pH of the medium was high, and so periodate is unlikely to be present as HIO_6^{4-}, as it is clear from its existence in multiple equilibria depending on the pH of the solution employed [16].

$$H_5IO_6 \underset{}{\overset{K_1}{\rightleftharpoons}} H_4IO_6^- + H^+ \qquad K_1 = 5.1 \times 10^{-4} \tag{2}$$

$$H_4IO_6^- \underset{}{\overset{K_2}{\rightleftharpoons}} H_3IO_6^{2-} + H^+ \qquad K_2 = 4.9 \times 10^{-9} \tag{3}$$

$$H_3IO_6^{2-} \underset{}{\overset{K_3}{\rightleftharpoons}} H_2IO_6^{3-} + H^+ \qquad K_3 = 2.5 \times 10^{-12} \tag{4}$$

In an acid medium, the periodic acid exists as H_5IO_6 and as $H_4IO_6^-$, but under the conditions employed in this study, i.e., in an alkaline medium, $H_3IO_6^{2-}$ and $H_2IO_6^{3-}$ are the main species expected. Periodate tends to dimerize if concentration is high [17], but the formation of this dimerized species is ruled out in the present kinetic conditions. Hence, at high pH used in this study, the water-soluble Copper (III) periodate complex is present as DPC, $[Cu(H_3IO_6)(H_2IO_6)]^{2-}$, as reported in earlier work [18, 19].

The oxidation reaction of MET with DPC in an alkaline medium has the stoichiometry 1:6 (MET:DPC) which is first-order dependent on [DPC] and an apparent less than unit order with respect to [MET], $[OH^-]$ and negative fractional order dependence on $[IO_4^-]$. The initially added products showed no effect on the oxidation reaction. Based on the experimental outcomes, a suitable mechanism is proposed for the oxidation of MET which has well accommodated with all the observed orders in each ingredient such as [oxidant], [reductant], [alkali], and [periodate]. The Copper (III) periodate complex in an alkaline medium exists as diperiodatocuprate (III)(DPC), monoperiodatocuprate (III)(MPC), and tertahydroxocuprate (III) as proposed by Lister [20]. In this kinetic study, DPC and MPC are considered active forms of Copper (III) periodate complex, since the existence of tertahydroxocuprate (III) was ruled out as its equilibrium constant is 8.0 x 10^{-11} at 40°C. As there is an increase in the rate of reaction with an increase in alkali concentration and a decrease in reaction rate with the increase in periodate concentration, suggesting the following possible equilibria of different Copper (III) periodate complexes as in Equations 5 and 6.

$$[Cu(H_3IO_6)_2]^- + OH^- \underset{}{\overset{K_4}{\rightleftharpoons}} [Cu(H_2IO_6)(H_3IO_6)]^{2-} + H_2O \quad (5)$$

$$[Cu(H_2IO_6)(H_3IO_6)]^{2-} + 2H_2O \underset{}{\overset{K_5}{\rightleftharpoons}} [Cu(H_2IO_6)(H_2O)_2] + [H_3IO_6]^{2-} \quad (6)$$

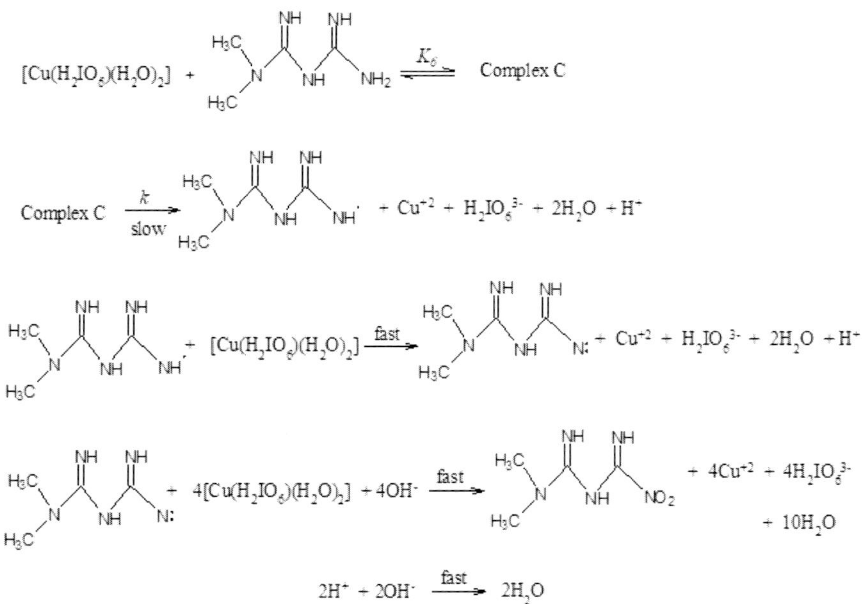

Scheme 1. Proposed mechanism for the oxidation of metformin by Cu(III) periodate complex.

The less than unit order dependence with respect to [MET] indicated that, there is formation of a Complex (C) between the oxidant, DPC and MET prior to the formation of the products. K_6 is the combined equilibrium constant including the equilibrium to bind active species of MET to MPC species which results in the formation of a Complex (C). The Complex (C) so formed decomposes in a rate-determining step to form a free radical of MET, Cu (II), and a water molecule, which is followed by consecutive fast steps to yield the reaction products as shown in Scheme 1.

This scheme is in agreement with the oxidation reactions taking place in the succession of one-electron steps, which is a well-understood principle of non-complimentary oxidation reactions. This type of reaction between a substrate and an oxidant would afford a radical intermediate, which is confirmed by the free radical scavenging experiment. This has also been observed and reported in earlier works [21, 22].

The formation of a complex between oxidant and substrate was confirmed by recording UV-Vis spectra of MET (1.0×10^{-4} mol dm^{-3}), DPC (1.0×10^{-5} mol dm^{-3}), [OH$^-$] = 0.04 mol dm^{-3}) and a mixture of both. A bathochromic shift of about 6 nm from 233 to 239 nm in the spectra of MET to a mixture of DPC and MET was observed. The complex formation between MET and DPC was also confirmed by the Michaelis-Menten plot, which is making it clear the less than unit order dependence of [MET]. Such a type of complex formation among a substrate and an oxidant was detected elsewhere [23, 24].

According to Scheme 1, the rate law can be derived as:

$$\text{Rate} = \frac{-d[\text{DPC}]}{dt} = k\,[\text{C}] = \frac{kK_4K_5K_6[\text{DPC}][\text{OH}^-][\text{MET}]}{[\text{H}_3\text{IO}_6^{2-}]} \tag{7}$$

$$[\text{DPC}]_T = [\text{DPC}]_f + [\text{Cu}(\text{H}_2\text{IO}_6)(\text{H}_3\text{IO}_6)^{2-}] + [\text{Cu}(\text{H}_2\text{IO}_6)(\text{H}_2\text{O})_2] + [\text{C}]$$

$$= [\text{DPC}]_f\left\{1 + K_4[\text{OH}^-] + \frac{K_4K_5[\text{OH}^-]}{[\text{H}_3\text{IO}_6^{2-}]} + \frac{K_4K_5K_6[\text{OH}^-][\text{MET}]}{[\text{H}_3\text{IO}_6^{2-}]}\right\}$$

where the subscripts, T and f refer to the total and free concentrations. Therefore:

$$[DPC]_f = \frac{[DPC]_T[H_3IO_6^{2-}]}{[H_3IO_6^{2-}] + K_4[H_3IO_6^{2-}][OH^-] + K_4K_5[OH^-] + K_4K_5K_6[OH^-][MET]} \tag{8}$$

Likewise, $[MET]_T = [MET]_f + [C]$

$$[MET]_f = \frac{[MET]_T[H_3IO_6^{2-}]}{[H_3IO_6^{2-}] + K_4K_5K_6[OH^-][Cu(H_3IO_6)_2]^-}$$

Because of the low concentration of DPC employed, the term $K_4K_5K_6[OH^-][Cu(H_3IO_6)_2]^-$ compared to $[H_3IO_6^{2-}]$ can be neglected. Therefore, we can write:

$$[MET]_f \cong [MET]_T \tag{9}$$

Likewise,

$$[OH^-]_f \cong [OH^-]_T \tag{10}$$

By substituting equations (8), (9) and (10) in equation (7) and omitting the subscripts t and f, we obtain:

$$\text{Rate} = \frac{kK_4K_5K_6[DPC][MET][OH^-]}{[H_3IO_6^{2-}] + K_4[OH^-][H_3IO_6^{2-}] + K_4K_5[OH^-] + K_4K_5K_6[MET][OH^-]} \tag{11}$$

$$\frac{\text{Rate}}{[DPC]} = k_{obs} = \frac{kK_4K_5K_6[MET][OH^-]}{[H_3IO_6^{2-}] + K_4[OH^-][H_3IO_6^{2-}] + K_4K_5[OH^-] + K_4K_5K_6[MET][OH^-]} \tag{12}$$

which clarifies all the observed orders of various species involved during kinetic runs.

The above rate law (12) can be reshuffled to the following form which is appropriate for verification:

$$\frac{1}{k_{obs}} = \frac{[H_3IO_6^{2-}]}{kK_4K_5K_6[OH^-][MET]} + \frac{[H_3IO_6^{2-}]}{kK_4K_5K_6[MET]} + \frac{1}{kK_6[MET]} + \frac{1}{k} \tag{13}$$

According to the above Equation (13), by keeping other conditions constant, graphs of $1/k_{obs}$ against 1/[MET] ($r^2 = 0.994$), $1/k_{obs}$ against $1/[OH^-]$ ($r^2 = 0.995$), and $1/k_{obs}$ against $[H_3IO_6^{2-}]$ ($r^2 = 0.994$) should be linear and they were found to follow that behavior as shown in Figure 3, Figure 4 and Figure 5. From the slopes and intercepts of these graphs, the values of K_4, K_5, K_6 and k were computed to be (58.82 ± 0.4) $dm^3\ mol^{-1}$, (0.23 ± 0.02) x 10^{-4} mol dm^{-3}, (3.07 ± 0.3) x 10^3 $dm^3\ mol^{-1}$, and (7.07 ± 0.35) x 10^{-3} s^{-1}, respectively. From the K_4 ($dm^3\ mol^{-1}$), K_5 (mol dm^{-3}), K_6 ($dm^3\ mol^{-1}$), and k (s^{-1}) values, the rate constants under various experimental conditions were computed by equation (12) and compared with experimentally observed data. As seen in Table 1, there is good agreement among them, which strengthens Scheme 1.

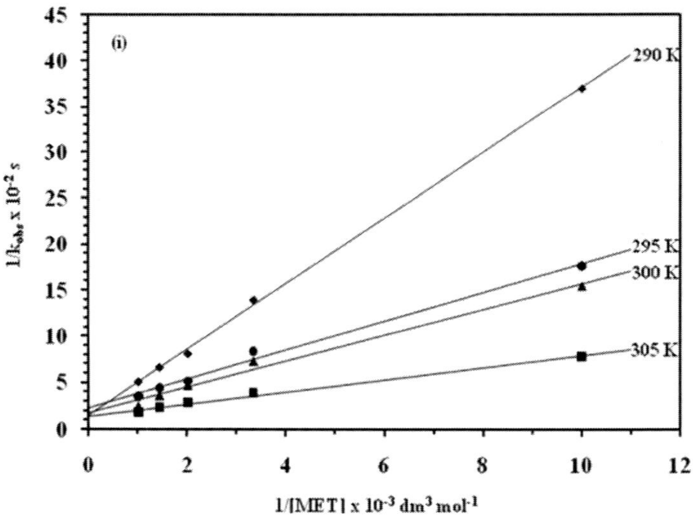

Figure 3. Graph of $1/k_{obs}$ versus 1/[MET] at four different temperatures (conditions as in Table 1).

Due to the better tendency of DPC to undergo hydrolysis compared to dissociation of hydrolyzed species in an aqueous alkaline medium, the equilibrium constant K_4 ($dm^3\ mol^{-1}$) is far bigger than K_5 (mol dm^{-3}). The influence of ionic strength on the reaction rate is not matching with proposed mechanism, which might be owing to the presence of high ionic strength present in this reaction. The dielectric constant has a negligible effect on the rate of reaction, which is opposite to the anticipated reaction between the ions having the identical charge in a media of lower relative permittivity.

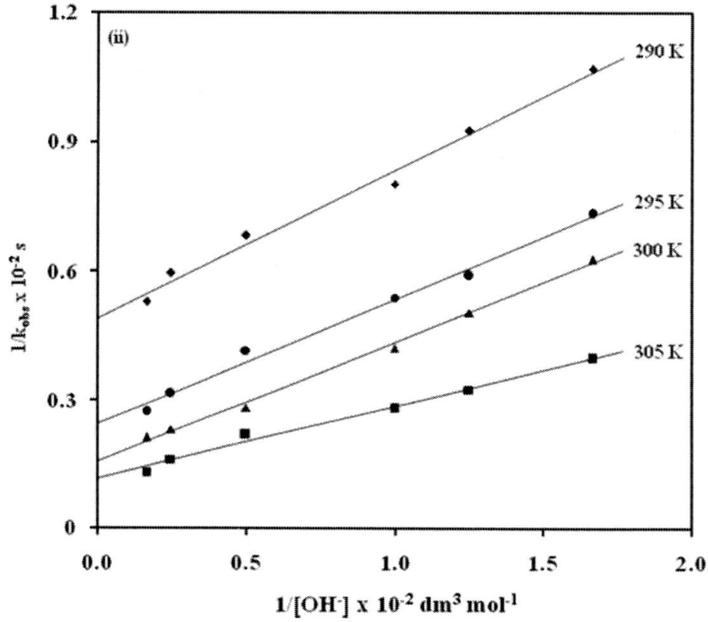

Figure 4. Graph of $1/k_{obs}$ versus $1/[OH^-]$ at four different temperatures (conditions as in Table 1).

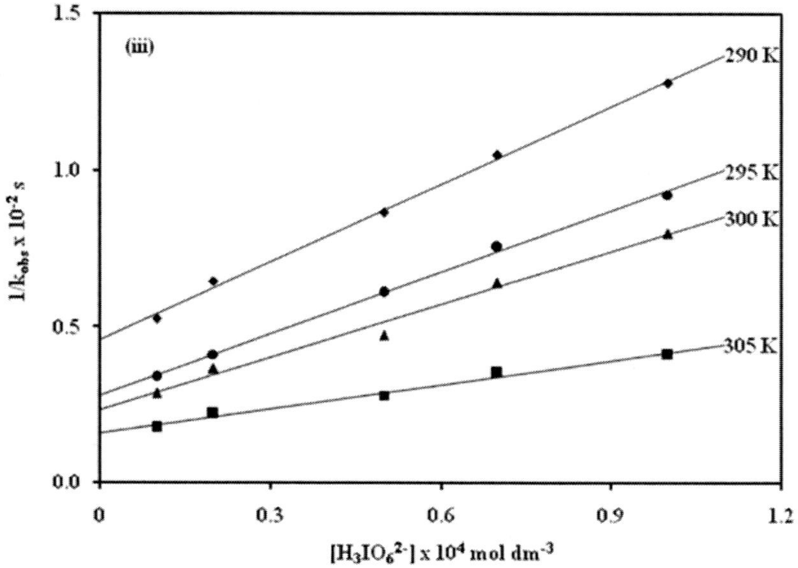

Figure 5. Graph of $1/k_{obs}$ versus $[H_3IO_6^{2-}]$ at four different temperatures (conditions as in Table 1).

To examine the effect of temperature, the kinetic runs were carried out at four different temperatures by taking constant quantities of DPC, MET, KOH and KNO_3. From the intercepts of the plot of $1/k_{obs}$ versus 1/[MET] ($r^2 \geq 0.994$), the rate constants, *k*, of the slow step of Scheme 1 were calculated at four different temperatures. Using the Arrhenius plot of $\log_{10}k$ versus 1/T, the energy of activation was computed, and from which the activation parameters were calculated and are shown in Table 2.

Table 2. Influence of temperature and activation parameters for the oxidation of MET with DPC in aqueous alkaline medium with respect to the rate determining step of Scheme 1

(A) Influence of temperature

Temperature (K)	k x 10^3 (s^{-1})
290	3.68
295	5.18
300	7.07
305	9.86

(B) Activation Parameters (Scheme 1)

Parameters	Values
Ea(kJ mol^{-1})	48 ± 1.2
$\Delta H^{\#}$ (kJ mol^{-1})	45 ± 0.8
$\Delta S^{\#}$ (JK^{-1} mol^{-1})	-134 ± 3
$\Delta G^{\#}$ (kJ mol^{-1})	85 ± 3
log A	6.2 ± 0.4

(C) Influence of temperature on the equilibrium constants of Scheme 1 for the oxidation of MET with diperiodatocuprate (III) in aqueous alkaline medium

Temperature (K)	K_4	K_5 x 10^4	K_6 x 10^{-3}
	(dm^3 mol^{-1})	(mol dm^{-3})	(dm^3 mol^{-1})
290	16.37 ± 0.3	0.99 ± 0.06	1.24 ± 0.3
295	28.47 ± 0.4	0.56 ± 0.05	2.19 ± 0.4
300	58.82 ± 0.4	0.23 ± 0.02	3.07 ± 0.3
305	97.53 ± 0.5	0.12 ± 0.02	5.38 ± 0.2

(D) Thermodynamic quantities utilizing the values of K_4, K_5 and K_6

Thermodynamic quantities	Values from	Values from	Values from
	K_4	K_5	K_6
Δ H (k J mol^{-1})	89 ± 1.2	-106 ± 3	70 ± 2
Δ S (J K^{-1} mol^{-1})	332 ± 2	-443 ± 4	301 ± 3
ΔG_{300} (k J mol^{-1})	-13 ± 0.9	31 ± 2	-23 ± 1

The thermodynamic quantities for the first, second, and third equilibrium steps of Scheme 1 were computed as follows. As shown in Table 1, the [MET], [$H_3IO_6^{2-}$] and [OH^-] were varied at four different temperatures. The graphs of $1/k_{obs}$ against 1/[MET], $1/k_{obs}$ against 1/[OH^-] and $1/k_{obs}$ against [$H_3IO_6^{2-}$] were plotted and found linear (Figure 3, Figure 4 and Figure 5). From the slopes and intercepts of these graphs, the K_4 ($dm^3\ mol^{-1}$) values were computed at four different temperatures and are tabulated in Table 2. A Van't Hoff's graph was plotted for the variation of K_4 ($dm^3\ mol^{-1}$) with temperature, i.e., $\log_{10}K_4$ against 1/T ($r^2 = 0.998$) and the values of enthalpy of reaction ΔH (k J mol^{-1}), entropy of reaction ΔS (J K^{-1} mol^{-1}), and free energy of reaction ΔG (k J mol^{-1}) were computed for the first equilibrium step (Table 2). In the similar way, the values of K_5 (mol dm^{-3}) and K_6 ($dm^3\ mol^{-1}$) were evaluated at different temperatures and the corresponding thermodynamic quantities were listed in Table 2.

The values of $\Delta H^{\#}$ (kJ mol^{-1}) and $\Delta S^{\#}$ (J K^{-1} mol^{-1}) indicate a favorable condition for an electron transfer process. The negative value of $\Delta S^{\#}$ (J K^{-1} mol^{-1}) shows that the complex (C) is more ordered than the reactants themselves [25]. A low quantity of the entropy of activation and a moderate value of the enthalpy of activation and a higher value of the rate constant for the slow step points out that the oxidation reaction takes place most probably via the inner-sphere mechanism. The earlier reported works support this fact [26, 27].

Conclusion

In this study, [$Cu(H_2IO_6)(H_2O)_2$] was believed to be the active species even though various species of DPC do exist in an aqueous alkaline medium. The title oxidation reaction was pH dependent. The rate constant for the rate-determining step and other equilibrium constants of the mechanism were computed. The product of the oxidation of metformin was confirmed by a spectral study. The influence of temperature was studied, and activation parameters were computed. The overall mechanistic study explained here is consistent with product, spectral and kinetic observations.

References

[1] Kumar, A., Kumar, P., Ramamurthy, P., (1999). Kinetics of oxidation of glycine and related substrates by diperiodatoargentate (III). *Polyhedron* 18, 773-780.

[2] Konnur, S. B., Nandibewoor, S. T., (2019). Kinetic investigations of ruthenium (III) catalyzed oxidation of dimedone by diperiodatocuprate(III) in aqueous alkaline media. *Rus. J. Phy. Chem.* A 93, 1686-1698.

[3] Shan, H., Qian, J., Gao, M. Z., Shen, S. G. Sun, H. W., (2004). Kinetics and mechanism of oxidation of n-propanolamine by dihydroxydiperiodatonickelate(IV) in alkaline medium. *Turk. J. Chem.* 28, 9-15.

[4] Niu, W., Zhu, Y., Hu, K., Tong, C., Yang, H., (1996). Kinetics of oxidation of SCN^- by diperiodatocuprate(III) in alkaline medium. *Int. J. Chem. Kin.* 28, 899-904.

[5] Karlin, K. D., Kaderli, S., Zuberbühler, A. D., (1997). Kinetics and thermodynamics of copper(I)/dioxygen interaction. *Acc. Chem. Res.* 30, 139-147.

[6] Sethuram, B., *Some Aspects of Electron Transfer Reactions Involving Organic Molecules.* Allied Publishers, New Delhi, 2003.

[7] Bailar, J. C., Emeleus, H. J., Nyholm, S. R., Trotman-Dickenson, A. F., *Comprehensive Inorganic Chemistry. Vol. 2*, Oxford, Pergamon, 1975.

[8] Setter, S. M., Iltz, J. L., Thams, J., Campbell, R. K., (2003). Metformin hydrochloride in the treatment of type 2 diabetes mellitus: a clinical review with a focus on dual therapy. *Clin. Therap.* 25, 2991-3026.

[9] Yerevanian, A., Soukas, A. A., (2019). Metformin: mechanisms in human obesity and weight loss. *Curr. Obes. Rep.* 8, 156-164.

[10] Ruggiero-Lopez, D., Lecomte, M., Moinet, G., Patereau, G., Lagarde, M., Wiernsperger, N., (1999). Reaction of metformin with dicarbonyl compounds possible implication in the inhibition of advanced glycation end product formation. *Biochem. Pharmac.* 58, 1765-1773.

[11] Gnesin, F., Thuesen, A. C., Kahler, L. K., Gluud, C., Madsbad, S., Hemmingsen, B., (2018). Metformin monotherapy for adults with type 2 diabetes mellitus. *Cochrane Database Syst. Rev.* 1, CD012906.

[12] Murthy, C. P., Sethuram, B., Navaneeth Rao, T., (1981). Kinetics of oxidation of some alcohols by diperiodatocuprate(III) in alkaline medium. *Z. Phys. Chem.* 262, 336-340.

[13] Jeffery, G. H., Bassett, J., Mendham, J., Denney, R. C., *Vogel's Text Book of Quantitative Chemical Analysis. 5th ed*, Essex, ELBS Longman, 1996.

[14] Gowda, J. I., Sataradi, S. R., Nandibewoor, S. T., (2012). Oxidation of xylitol by a silver (III) periodate complex in the presence of osmium (VIII) as a homogeneous catalyst. *Catal. Sci. Technol.* 2, 2549-2557.

[15] Bhattacharya, S., Banerjee, P., (1996). Kinetic studies on the electron transfer between azide and nickel(V) oxime imine complexs in aqueous solution. *Bul. Chem. Soc. Japan,* 69, 3475-3482

[16] Gowda, J. I., Nayak, S. S., Langote, S. R., Joshi, P. S., Nandibewoor, S. T., (2015). Spectroscopic and mechanistic investigation into oxidation of aspartame by diperiodatocuprate(III) in aqueous alkaline medium. *Cogent Chem.* 1, 1015909.

[17] Sethuram, B. *Some Aspects of Electron Transfer Reactions Involving Organic Molecules*. Allied Publishers, New Delhi, 2003.

[18] Hegde, R. N., Shetti, N. P., Nandibewoor, S. T., (2009). Oxidative degradation and deamination of atenolol by diperiodatocuprate (III) in aqueous alkaline medium: A mechanistic study. *Polyhedron* 28, 3499-3506.

[19] Murthy, C. P., Sethuram, B., Reddy, K. B., Rao, T. N., (1984). Kinetics of oxidation of benzaldehydes by copper (III) in t-butanol water medium. *Indian J. Chem.* 23A, 593-595.

[20] Lister, M. W., (1953). The stability of some complexes of trivalent copper. *Can. J. Chem.* 31, 638-652.

[21] Jaky, M., Szeverenyi, M., Simandi, L. I., (1991). Formation of manganate(V) in oxidations by permanganate ion in strongly alkaline solutions. *Inorg. Chim. Acta* 186, 33-37.

[22] Hosahalli, R. V., Byadgi, K. S., Nandibewoor, S. T., Chimatadar, S. A., (2010). Kinetics and mechanism of oxidation of chloramphenicol-an antibiotic drug by diperiodatocuprate(III) in aqueous alkaline medium. *Z. Phys. Chem.* 225, 79-94.

[23] Bilehal, D. C., Kulkarni, R. M. Nandibewoor, S. T., (2003). Oxidation of nicotinate ion by heptavalent manganese: a kinetic and mechanistic approach. *Z. Phys. Chem.* 217, 1-12.

[24] Thimmegowda, B., Ishwarabhat, J., (1989). Kinetics of oxidation of thiocyanate ion by n-chloro and n-bromosuccinimides in aqueous methanol and alkaline media. *Indian J. Chem.* 28A, 43-46.

[25] Weissberger, A., Lewis, E. S. *Investigation of Rates and Mechanism of Reactions in Techniques of Chemistry.* Wiley Interscience, New York, 1974.

[26] Martinez, M., Pitarque, M. A. Eldik, R. V. (1996). Outer-sphere *RedOx* reactions of $[Co^{III}(NH_3)_5(H_xP_yO_z)]^{(m-3)-}$ complexes. A temperature- and pressure-dependence kinetic study on the influence of the phosphorus oxoanions. *J. Chem. Soc., Dalton Trans.* 2665-2671.

[27] Farokhi, S. A., Nandibewoor, S. T., (2003). Kinetic, mechanistic, and spectral studies for the oxidation of sulfanilic acid by alkaline hexacyoanoferrate (III). *Tetrahedron* 59, 7595-7602.

Chapter 7

Kinetics of Electrocatalytic Reactions

C. F. Zinola*
Electrochemical Engineering Group,
School of Sciences, School of Engineering,
Universidad de la República, Montevideo, Uruguay

Abstract

The Electrocatalytic Processes are almost 90% of the heterogeneous reactions since they are those affected by the nature of the substrate and the presence of an adsorbate (either as reactant, intermediate, or product). As these reactions are not sufficiently well studied, special emphasis is placed on the possibilities of enhancement or inhibition by properly designing their kinetic parameters. They can help to the manipulate the mechanism and thus obtain the desired products or simply enhancing the overall kinetics.

For the above purposes, this chapter covers the fundamentals of Electrode Kinetics, firstly considering the most important parameters: the symmetry factor and the exchange current density. Secondly, presenting the quantum nature of electron tunneling and transfer with a little history on the matter. Thirdly, analyzing the influence of the electrosorption processes in the activation energy curves to understand the possibilities of changing the rate determining step.

Since electrochemical reactions result in electrical currents, charge densities, and electrode potentials, their rates and extents are here defined to measure, control and design a selected process. The relative new experimental hybrid techniques are here exemplified to characterize their physicochemical aspects as well as the applications in a balanced style. Furthermore, this chapter serves as a textbook for professionals and

---

* Corresponding Author's Email: fzinola@fcien.edu.uy.

In: A Closer Look at Chemical Kinetics
Editor: Victor Martinez-Luaces
ISBN: 979-8-88697-484-3

scientists working in electrochemical industrial technologies and in research at educational centers or research institutions in various related fields.

The chapter is divided into three sections: (i) the fundamentals of Electrode Kinetics; (ii) the properties of Electrocatalysis; and (iii) the description of one electrocatalytic process with applications in fuel cells: methanol oxidation and its elimination.

Keywords: electrocatalysis, electrodes, kinetics, current density, overpotential

Introduction: Electrode Kinetics

Tafel Law at the Electrochemical Interface

The creation of Electrode Kinetics dated 1905 with Julius Tafel's work. An empirical law was derived based on experimental data, unaccepted by physicists since it denied Ohm's Law. It is also the basement of Electrocatalysis since it involved the combination of an electric field at the interface and the acceleration of reactions by a change in substrate.

Tafel's Law is [1] defined as:

$$\eta = a \pm b \log j \tag{1}$$

The Tafel parameters being being *a* and *b*. The first one involves reversible components and the second one is the so-called Tafel slope. The functions involved are: η the overpotential (the difference between the electrode potential and its reversible value) for the electrode reaction and j the current density developing at the electrified interface. The current "density" refers to the standardization with respect to the real or geometric area under consideration.

At suitable large values of η_{linear}, a linear region with the logarithm of j is observed (Figure 1). The extrapolation of this line at $\eta = 0$ gives a characteristic kinetic property named *exchange current density*, j_o (defined with a and b).

In electrochemical phenomena, and hence Electrocatalysis, the terminology is quite different from Chemical Kinetics. In Chemical Catalysis the measurement of a reaction rate is in *mol per square centimeter per second*, since it is the way to evaluate the reaction rate. In electrocatalysis, the measurement of the rate is in current density units. Applying Faraday's Law we realized that j/nF means the number of times the overall reaction has occurred; and that its units also means mol cm^{-2} s^{-1}. Thus, electrocatalytic data can be converted to the catalytic nomenclature and *vice versa.*

Tafel's law limitations arise at sufficiently high j values since less amounts of reactants are available for the electrode reaction (at a given η value). As a result, a limiting current density, j_L is attained and the Tafel's Law will be not appropriate to describe the process. Thus, a time-dependent local concentration (at the surface) has to be considered, which is a function of η.

Anyway, working at moderate values of j and η, we can conveniently re-write the law in its exponential form:

$$j = j_o \exp(\pm \alpha F \eta / RT) \tag{2}$$

where j_o is an intrinsic minimum value of j for the electrode reaction at zero η, α is the *electronic charge transfer coefficient* (defined as the part of the electric energy affecting the activation Gibbs energy), F is the Faraday constant (96500 C mol^{-1}), R is the Regnault constant (8.314 J mol^{-1} K^{-1}) and T is the absolute temperature.

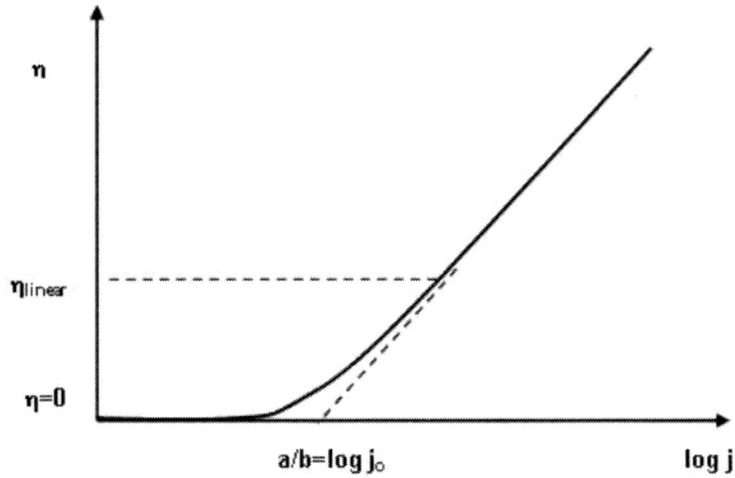

Figure 1. Tafel's semi-logarithmic line for a single electron transfer reaction. The exponential relationship under the high field approximation tends to a linear overpotential (η) vs. logarithm current density (j) expression.

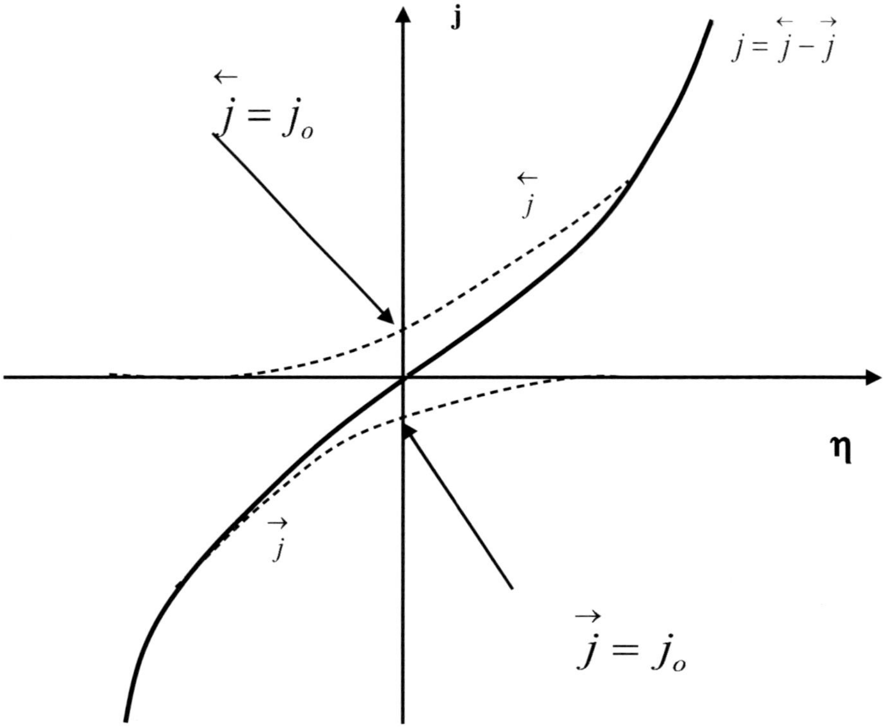

Figure 2. The current density (j) *vs.* overpotential (η) curve with the exchange current density (j_o) at zero overpotential and null current density.

There are two exponential vectorial components of the entire reaction obeying a Tafel law; one for the reduction reaction $\vec{j}$ and another one for the oxidation reaction $\overleftarrow{j}$. The resulting current density, j, has to be positive, so it is simply the difference between both of them. When the interface is able to develop an invertible reaction, we can plot the complete perturbation in 4 quadrants (Figure 2). The first one is assigned to the oxidation reaction (at the anode) and the third to the reduction reaction (at the cathode). According to the Faraday's nomenclature, a *cathodic reaction* means a reaction in which the electrons leave the electrode for an entity in the double layer next to the electrode (adsorbed or not). However, an *anodic reaction* indicates a reaction in which the reactants leave the electrode or the double layer, yielding a higher oxidation number (adsorbed) species as product (leaving electrons on the surface).

Electron Distribution at the Electrified Interface

Since the electron transport and stability at the different zones of the electrified interface are the basis of Electrode Kinetics, we have to treat the topic using Quantum Mechanics. *Electron distributions* are governed by the so-called *Density of States* (the "zone" including atoms with the same energy). The simplest case is described by the *Fermi distribution* with non-interactive and attractive or repulsive statistics. Thus, the application of the *Pauli Exclusion Principle* implies that the occupancy of any permitted state is restricted to 1 or 0 electrons. However, Fermi's law gives the probability of setting a given number of electrons, *P*, at a certain energy state, *E*, occupied in the metal, i.e., the *Fermi-Dirac Distribution*:

$$P = \frac{1}{1 + \exp\left[\frac{E - E_f}{kT}\right]} \tag{3}$$

being E_f the Fermi Energy of the conduction electrons in the metal taken as the maximum kinetic energy from its base at T → 0. Thus, at ordinary temperatures, most of the energy levels are occupied up to E_f and a little amount of electrons has energy above it. For copper, this level is 7 eV, for silver 5.49 eV and gold 5.53 eV, whereas the thermic energy, *kT*, is only 0.025 at 300 K. If those numbers are introduced in Eq. (3), its value is nearly 1 up to E_f and quickly reaches 0 above it [2].

The density of states, $\rho(E)$, gives the number of electrons available to create a current density between the energy *E* and *E* + *dE*:

$$\rho(E) = \frac{1}{2\pi^2}\left(\frac{2m}{\hbar^2}\right)^{3/2} E^{1/2} \tag{4}$$

Multiplying by *P*, the conduction electron population can be obtained by integrating the total energy *E* between 0 and ∞ [3].

$$n = \int_0^\infty \rho(E)P(E)dE = \frac{1}{2\pi^2}\left(\frac{2m}{\hbar^2}\right)^{3/2} \int_0^\infty \frac{E^{1/2}dE}{\exp\left[(E - E_f)/kT\right] + 1} \tag{5}$$

The result is given by Jonquière's function, *Li*;

$$n(E_f) = -\frac{\sqrt{\pi}}{4\pi^2}\left(\frac{2kTm}{\hbar^2}\right)^{3/2} Li_{3/2}\left(-e^{E_f/kT}\right) \tag{6}$$

where *Li* is the *poly-logarithmic integral function* defined as a series combination (Euler sums) in the complex plane.

The solution for real physical parameters is also written as:

$$n(E_f) = \frac{1}{\sqrt{2}}\left(\frac{kTm}{\pi\hbar^2}\right)^{3/2} e^{E_f/kT} Li_2(3/2) \tag{7}$$

For the example given above for E_f at 300 K, the obtained electronic population is: $n = 7.43\ 10^{114} - (3.99\ 10^{114})\ i$. However, at 0 K the value of n decreases to 10^{29} electrons m^{-3}.

A similar integration is achieved taking the infinity as E_f at 0 K, so:

$$n(E_f) = \frac{1}{\sqrt{2}}\left(\frac{kTm}{\pi\hbar^2}\right)^{3/2} \frac{2}{3} E_f^{\ 3/2} \tag{8}$$

Then, the E_f is obtained as a function of the electronic population:

$$E_f = \frac{h^2}{8m}\left(\frac{3}{\pi}\right)^{2/3} n^{2/3} \tag{9}$$

The real part of this function can be plotted for the occupied density of states $\rho(E)$, as a function of energy, E (Figure 3). The value of E_f at T = 0 K is moved to lower values when the temperature increases.

The current density is the result of a change in the occupancy of the states near E_f, which separates the unfilled orbitals in the metal from the filled ones in the linear momentum space; $\bar{p} = \hbar\ \bar{k}$, where $\bar{k}$ is the wave vector number, that is defined in a real space of cartesian coordinates as: $\bar{k} = [k_x, k_y, k_z]$.

The energy is modified by 2 components. First, the electrostatic energy, E_e, as the product of the charge density, $\rho(E)$, and the electric potential, $V_e(r)$, of the valence electrons.

$$E_e = 1/2\int_0^{r_1} \rho(r)V_e(r)d^3r - c\int_0^{r_1}\left[1+\frac{c^`}{1+c^{``}[\rho(r)]^{1/3}}\right][\rho(r)]^{1/3}d^3r \qquad (10)$$

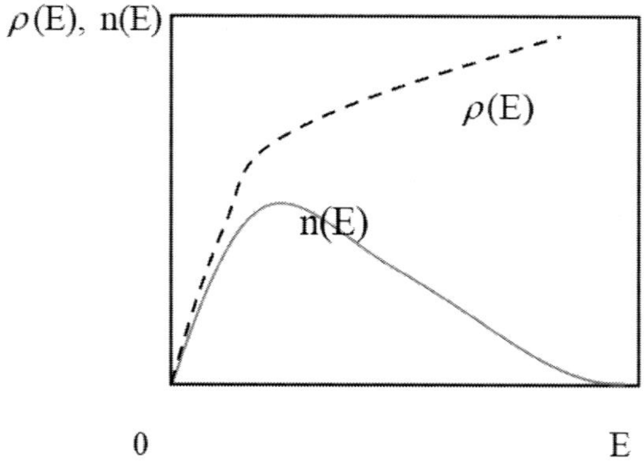

Figure 3. Absolute Density of States, $\rho(E)$, and the electron population, $n(E)$, vs. Energy, E.

And second the core energy, E_c, as the product of $\rho(E)$ with the core electric potential, V_c.

$$E_c = \int_0^{r_1} \rho(r)V_c(r)d^3r \qquad (11)$$

A good arrangement was found for liquid metals considering the *jellium model*. On the other hand, for a solid, the criteria do not give good results and more corrections for the interaction energies are needed [4].

The surface energy is taken as the contribution of the bulk of the metal and that of the surface. The separation between these two spaces is the surface energy, which is strongly dependent on the crystallographic order and its relaxation. However, the energy space that counts to define the current density is the kinetic energy of the valence electrons, E_k (next page):

$$E_k = \sum_i \int_0^{r_1} \varphi_i^*(r)(-1/2\nabla^2)\varphi_i(r)d^3r \tag{12}$$

where ½ ∇^2 is the kinetic energy operator, $\varphi_i(r)$ and $\varphi_i^*(r)$ the real and complex atomic conjugated orbitals and r_1 the radius of the bulk metal.

The surface states are defined by the *Linear Combination Atomic Orbital* (LCAO) method. However, it is better to evaluate them through the electrode potential and distance from the electrode surface relationship. These electric potentials obey the *Schrödinger equation* and the surface states are obtained as eigenvalues. In any case, using the *LCAO* method, the interaction between the surface states and any electron is avoided and the suggested wave function Ψ_α, is:

$$\Psi_\alpha(r) = \sum_i a_{\alpha,i}\varphi_i(r) \tag{13}$$

where $a_{\alpha,i}$ are the linear independent expansion coefficients. The proper choice of $\varphi_i(r)$ (base functions) is the right clue for obtaining Ψ_α in a metal lattice. In this case, *Bloch functions* [2] are the right choice for the metal atomic orbitals. The Bloch Theorem formulates a condition that must be satisfied by all solutions, $\Psi_\alpha(r)$, for any periodic potential, $V(r)$, at which the electrode is submitted to.

It ascertains:

$$\Psi(r) = u(r)\exp(ik.r) \tag{14}$$

where $u(r)$ is any function that exhibits a periodicity in the lattice, i.e., $u(r)= u(r+r_o)$.

Each $\Psi(r)$ that accomplishes this requirement is called a *Bloch wave*. In fact, the expansion of $V(r)$ and $\Psi(r)$ into Fourier Series possibly this equation:

$$\Psi(r + r_o) = \Psi(r)\exp(ik.r_o) \tag{15}$$

It implies that any function $\Psi(r)$ that is a solution to the Schrödinger equation of the metal lattice varies only by a phase factor $exp(i\ k.r_o)$ between adjacent and equivalent positions in the metal lattice. In other words, the probability of finding an electron at any equivalent position of the crystal lattice is the same.

The associated energy can be obtained as a mean value:

$$\langle E \rangle = \frac{\int \Psi^* H \Psi d\tau}{\int \Psi^* \Psi d\tau} = \frac{\int \Psi^* H \Psi d\tau}{\int |\Psi|^2 d\tau} \tag{16}$$

An exact solution is not always obtained and a first approximation of zero contribution for overlapping integrals leads to only coulombic and resonance integrals for the nearest neighbors. The difference between the surface and inner coulomb integrals produces the supposed "deformation" [5]. The solution of these integrals gives the surface states.

In the case of an electric field at the interface, the wave functions are modifications of Bloch functions. The latter develop an orthogonal set whose components are completely determined, but not the phase magnitudes. The theory displays them as power series in the field strength; the unchanged Bloch band functions form the zero-order term of that series [6].

Many papers have been published on the rate of electrode reactions, but a few on the quantum approach of electronic transfer at the interface. It has been observed that the electrostatic model of the Gibbs Energy of Activation cannot predict the linearity of a Tafel line and also the constancy of the symmetry factor. The thermal approximation is better; however, the relationship between the current density and the electrode potential perturbation is given by [7]:

$$j = nFC_{x=0} \int_0^{E_f} \int_0^{x} n(E) \rho(E) W_{E,x}(E) dx dE \tag{17}$$

$C_{x=0}$ being the surface concentration of the reactant and $W_{E,x}$ the transition probability, which is the main difference in the calculation for the selected method.

A quantum mechanical derivation can be given to show the consistency of the time-dependent perturbation theory with the *Gamow Equation* for the time-dependent barrier concerned in charge transfer [7].

However, for a constant (square barrier) potential, the Gamow Equation gives a similar solution to the Wentzel, Kramer and Brillouin (*WKB)* approach, as the following is stated.

Electron Tunneling at the Electrochemical Interface

The Schrödinger equation of an electron at the electrochemical interface taking the metal as a potential barrier can be elucidated considering that the electron kinetic energy is less than the potential barrier. For simplicity, only the 1-D approach is presented using the Eckart potential energy *U(x)* profile [8]. The reaction coordinate, *x*, is normalized by the barrier width *2L* to the function $\xi \equiv 2\pi x/L$.

When *U(x)* is larger than the total energy of the electron, *E*:

$$U(\xi) = \frac{\Delta H_{reac} e^{\xi}}{1+e^{\xi}} + \frac{\Delta H_{act} e^{\xi}}{\left(1+e^{\xi}\right)^2} \tag{18}$$

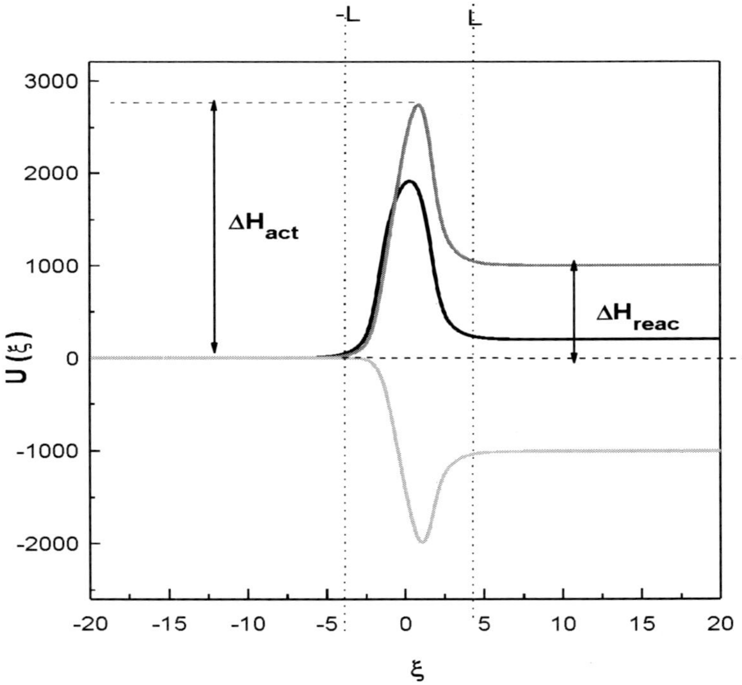

Figure 4. Eckart potential energy, $U(\xi)$, for different values of $\Delta \bar{H}_{reac}$ and $\Delta \bar{H}_{act}$ as a function of ξ (grey lines) $\Delta \bar{H}_{reac} = -1000$, $\Delta \bar{H}_{act} = +3000$; (black lines) $\Delta \bar{H}_{reac} = -200$; $\Delta \bar{H}_{act} = +2000$; (pale grey lines) $\Delta \bar{H}_{reac} = +1000$; $\Delta \bar{H}_{act} = -2000$ in arbitrary units.

Considering the electron of mass *m* moving in the potential barrier, the potential energy, $U(x)$ is zero for $x < -L$ and $x > L$. Now, outside the barrier it happens that:

$$\frac{-\hbar^2}{2m}\frac{d^2\psi(\xi)}{d\xi^2} = E\psi(\xi) \tag{19}$$

But inside the barrier the following situation occurs:

$$\frac{-\hbar^2}{2m}\frac{d^2\psi(\xi)}{d\xi^2} = [U(\xi) - E]\psi(\xi) \tag{20}$$

As a first approach, we can take a constant barrier $U(\xi) = U_o$ [9] to envisage the problem.

The solutions are then, a linear combination of exponential terms for $E < U_o$:

$$\psi(\xi) = Ae^{k\xi} + Be^{-k\xi} \qquad \text{With } k = \sqrt{\frac{2m(U_o - E)}{\hbar^2}} \tag{21}$$

The probability of electron transmission can be calculated from:

$$P = \frac{|\psi_{outside}(\xi)|^2}{|\psi_{incid}(\xi)|^2} \tag{22}$$

with $\Psi_i(\xi)$ being the wave functions of the electron at the outside and incidence side of the potential barrier.

Moreover, the solution given in Eq. (20) will be similar on both sides of the barrier but only the positive exponential terms are able to signify the transmission effects:

$$\psi_{incid}(\xi) = Ce^{ik'\xi} \qquad \text{With } k' = \sqrt{\frac{2mE}{\hbar^2}} \qquad (23a,b,c)$$
$$\psi_{transm}(\xi) = Ee^{ik'\xi}$$

Therefore, we obtain:

$$P = \left[\frac{16E}{U_o} - \left(\frac{4E}{U_o} \right)^2 \right] \exp(-2kL) \tag{24}$$

Employing Eq. (17) we can calculate *j* after considering in the reaction the reactant transition probability. The latter is assigned to a Maxwell-Boltzmann Distribution, so for a single *A* species at the interface yielding *B*, we say:

$$P_{AB}(E) = const \exp(-\Delta\mu_{AB}^{o} / kT) \tag{25}$$

Being $\Delta\mu_{AB}{}^{o}$ the change in the chemical energy for the reaction *A* ->*B*.

When there is an electric potential applied to the interface, $\Delta\phi$, the fraction of the activation energy modified by this electric term has to be; $\alpha\Delta\phi$. α is the electronic charge transfer coefficient, which in terms of an elementary reaction is only the symmetry factor of the reaction, β. In terms of energies, we have to multiply it with the elementary electronic charge, $e\alpha\Delta\phi$. Thus,

$$P_{AB}(E) = const \exp(-\Delta\mu_{AB}^{o} / kT) \exp(-e\beta\Delta\phi / kT) \tag{26}$$

Then, *j* will be in a simple "thermal" model the product between the electron probability, Eq. (24), and that of the reactant, Eq. (26):

$$j = nFC_{x=0} const \left[\frac{16E}{U_o} - \left(\frac{4E}{U_o} \right)^2 \right] \exp(-2kL) \exp(-\Delta\mu_{AB}^{o} / kT) \exp(-e\beta\Delta\phi / kT) \tag{27}$$

Moreover, taking the overpotential, η, as the difference between the electric potential and that at reversible conditions, $\eta \equiv \Delta\phi - \Delta\phi_{rev}$ and after substituting in Eq.(27), we can obtain:

$$j = nFC_{x=0} const \left[\frac{16E}{U_o} - \left(\frac{4E}{U_o} \right)^2 \right] \exp(-2kL) * \tag{28}$$

$$* \exp(-\Delta\mu_{AB}^{o} / kT) \exp(-e\beta\Delta\phi_{rev} / kT) \exp(-e\beta\eta / kT)$$

In other words, the Tafel's Law is proven again:

$$j = j_o \exp(-e\beta\eta / kT) \tag{29}$$

j_o being the exchange current density involving all the physicochemical parameters described in Eq. (28).

The WKB Approach to the Eckart Potential

When there is no constant barrier energy and the Eckart potential applies, it is possible to solve the problem by taking some approximations or by an exact analytical equation.

After substitution of Eq. (17) into (19) we can state that:

$$\frac{-\hbar^2}{2m}\frac{d^2\psi(\xi)}{d\xi^2} = \left(\frac{\Delta H_{reac} e^{\xi}}{1+e^{\xi}} + \frac{\Delta H_{act} e^{\xi}}{\left(1+e^{\xi}\right)^2} - E\right)\psi(\xi)$$

(30)

The WKB approach proposes a similar solution to (23) and (24):

$$\psi(\xi) = C(\xi)e^{ikS(\xi)} \tag{31}$$

where $S(\xi)$ has the wave vector number, $\bar{k}(\xi)$.

Substituting in Eq. (30) we will have:

$$C''(\xi) - C(\xi)[S'(\xi)]^2 - \frac{2m}{\hbar^2}(U(\xi) - E)C(\xi) + i[2C'(\xi)S'(\xi) + C(\xi)S''(\xi)] = 0$$

(32)

Both real and imaginary contributions have to annihilate separately:

$$C^2(\xi)S''(\xi) + 2C(\xi)C'(\xi)S'(\xi) = 0$$
$$\frac{d}{d\xi}\left[C^2(\xi)S'(\xi)\right] = 0$$

(33a,b)

Since the differential of a constant is zero, $C^2(\xi)S'(\xi)$ has to be the constant defined as C_o.

$$C(\xi) = \sqrt{\frac{C_o}{S'(\xi)}} \tag{34}$$

The variation of $C(\xi)$ with ξ will be negligible since it includes the first derivative of $S(\xi)$. Thus, we can surely state that $C''(\xi) \rightarrow 0$, being the *WKB* approach.

The wave function under the *WKB* approximation will be [10]:

$$\psi(\xi) = \frac{\sqrt{\frac{C_o}{i}} \exp\left\{\pm \int \sqrt{\frac{2m}{\hbar^2}(U(\xi) - E)} d\xi \right\}}{\left[\frac{2m}{\hbar^2}(U(\xi) - E)\right]^{1/4}} \tag{35}$$

with $E < U(\xi)$.

It can be demonstrated that this approximation accounts for $\lambda \cong \Delta x$, that is, when the particle wavelength is similar to the barrier width.

The Exact Solution to the Eckart Potential

The exact solution is obtained by the hypergeometric Gauss functions. In this treatment, another variable was taken into account, that is, the crystal wave spreading through the metal lattice. Thus, we can define the variable associated as $y = -\exp(2\pi x/L)$ for the Hamiltonian expression:

$$\frac{d^2\psi(x,y)}{dx^2} - \frac{2m}{\hbar^2}\left(\frac{\Delta H_{reac} y}{1-y} + \frac{\Delta H_{act} y}{(1-y)^2} - E\right)\psi(x,y) = 0 \tag{36}$$

but,

$$\frac{d^2\psi(x,y)}{dx^2} = \frac{d^2\psi(x,y)}{dy^2}\left(\frac{dy}{dx}\right)^2 + \frac{d\psi(x,y)}{dx}\left(\frac{d^2y}{dx^2}\right) \tag{37}$$

and then:

$$y^2\frac{d^2\psi(x,y)}{dy^2}+y\frac{d\psi(x,y)}{dy}+\frac{2mL}{\hbar^2}\left(E-\frac{\Delta H_{reac.}y}{1-y}+\frac{\Delta H_{act.}y}{(1-y)^2}\right)\psi(x,y)=0$$

(38)

In the case of the regions near $x\rightarrow\pm\infty$, there is no potential barrier, and the expression reduces to the well-known simple linear combination of exponentials.

However, for $-L<x<L$ (inside the potential energy barrier) the suggested solution is a product of two independent functions:

$$\psi(x,y)=\Omega(x)\Xi(y) \tag{39}$$

where $\Omega(x)$ is the asymptotic behavior of the wave function $\Psi(x,y)$ at $x\rightarrow\pm\infty$.

Under this condition $\Omega(x)$ becomes independent of x:

$$\Omega(y)=(1-y)^{i\beta}\left[\frac{y}{1-y}\right]^{i\alpha} \tag{40}$$

The other function is the solution of:

$$y(1-y)\frac{d^2\Xi(y)}{dy^2}+[c-(a+b+1)y]\frac{d\Xi(y)}{dy}-ab\Xi(y)=0 \tag{41}$$

with $ab\equiv\Delta H_{reac}-(\alpha-\beta)^2-i(\alpha-\beta)$, $c\equiv 1-2i\beta$ and $a+b+1=2+2i\alpha-2i\beta$ and also: $\alpha^2=E'$, $\beta^2=E'-\Delta H_{reac}$ and $\alpha^2-\beta^2=\Delta H_{reac}$.

The solution of $\Xi(y)$ is:

$$\Xi(y)=C_1F(a,b,c,y)+C_2y^{1-c}F(1+a-c+1+b-c+2-c-y)$$

(42)

where C_1 and C_2 are arbitrary constants and F is the Gauss hypergeometric function [11]. The polynomial expression of F obeys; $F(a,b,c,0)=1$.

For the general solution we suggest a linear combination of hypergeometric functions of the form:

$$F(a,b,c,y) = MF(a,b,a+b-c+1,1-y) + (1-y)^{c-a-b} NF(c-a,c-b,c-a-b+1,1-y) \tag{43}$$

being:

$$M = \frac{\Gamma(c)\Gamma(c-a-b)}{\Gamma(c-a)\Gamma(c-b)} \text{ and } N = \frac{\Gamma(c)\Gamma(a+b-c)}{\Gamma(a)\Gamma(b)} \tag{44a,b}$$

where *Γ(x)* is the Complete Gamma Function.

The function *F* is equal to *1* when $x \to \infty$ and $y \to 0$. The problem is the value of *F* when $x \to -\infty$ ($y \to 1$). In this case, it acquires the same figure and equal to 1 on both sides:

$$F(a,b,c,y) = M + (1-y)^{c-a-b} N = M + (1-y)^{-2i\alpha} N \tag{45}$$

And in the entire non-asymptotic solution, we will have:

$$\psi(x,y) = (1-y)^{i\beta} \left[\frac{y}{1-y}\right]^{i\alpha} My^{i\beta}(1-y)^{i\alpha} + Ny^{-i\beta}(1-y)^{-i\alpha} \tag{46}$$

For practical purposes we can define the transmission probability, *P*, as the complement of the reflection coefficient, *R*, as:

$$P = 1 - R = 1 - \frac{|a_2|^2}{|a_1|^2} = 1 - \frac{|N|^2}{|M|^2} \tag{47}$$

Taking Eq. (44) and substituting into Eq. (47) we have:

$$P = 1 - \left|\frac{\Gamma(a+b+c)\Gamma(c-b)\Gamma(c-a)}{\Gamma(a)\Gamma(b)\Gamma(c-a-b)}\right|^2 \tag{48}$$

But since $a + b - c = 2i\alpha$ and $c - a - b = - 2i\alpha$, it is possible to find that:

$$P = 1 - \left| \frac{\Gamma(c-b)\Gamma(c-a)}{\Gamma(a)\Gamma(b)} \right|^2 \tag{49}$$

with:

$$\begin{aligned} a &= 1/2 + i(\alpha - \beta + \gamma) \\ b &= 1/2 + i(\alpha - \beta - \gamma) \end{aligned} \tag{50a,b}$$

And since α and β are real and $\gamma \equiv$ ½ (4 B'-1)$^{1/2}$ we obtain:

$$P = 1 - \frac{ch\pi(\alpha - \beta + \gamma)ch\pi(\alpha - \beta - \gamma)}{ch\pi(\alpha + \beta + \gamma)ch\pi(\alpha + \beta - \gamma)} \tag{51}$$

Otherwise, we can reduce Eq. (51) using some combinations of the Moivre's formulae for hyperbolic trigonometry:

$$P = \frac{ch[2\pi(\alpha + \beta)] - ch[2\pi(\alpha - \beta)]}{ch[2\pi(\alpha + \beta)] + ch(2\pi\gamma)} \tag{52}$$

The above expression is only applicable when γ is real (B'> 1/4), where the wavelength λ is small compared to $2L$. In the case of being imaginary the method proposed in [12] has to be used.

Thus, the current density, j, for a hypothetical reaction with n electrons transferred at the interface will be the product of Eq. (60) with Eq. (26):

$$j = nFC_{x=0} \frac{ch[2\pi(\alpha + \beta)] - ch[2\pi(\alpha - \beta)]}{ch[2\pi(\alpha + \beta)] + ch(2\pi\gamma)} const \exp(-\Delta\mu_{AB}^{o} / kT) \exp(-e\beta\Delta\phi / kT) \tag{53}$$

β being the symmetry factor of the single elementary electrochemical path.

A similar equation can be derived in terms of overpotential, η:

$$j = nFC_{x=0} \frac{ch[2\pi(\alpha+\beta)] - ch[2\pi(\alpha-\beta)]}{ch[2\pi(\alpha+\beta)] + ch(2\pi\gamma)} const * \tag{54}$$
$$* \exp(-\Delta\mu^{o}_{AB} / kT) \exp(-e\beta\Delta\phi_{rev} / kT) \exp(-e\beta\eta / kT)$$

Or again our known Tafel's Law:

$$j = j_{o}^{*} \exp(-e\beta\eta / kT) \tag{55}$$

j_o^* being the new exchange current density condensing the physico-chemical parameters in Eq. (54).

The Charge Transfer across the Electrochemical Interface

The treatment of charge transfer through an energy barrier in the electronic/ionic interface was firstly proposed by Gurney in 1931 [13].

He considered the electron tunneling across an electrochemical interface from the metal ground states to an ion in solution. His model had a fault, which prevented Gurney's theory from being applied to give a basis to Electrocatalysis. Gurney proposed that the electrons leave the metal "neutralizing" H_3O^+ in the double layer. The result was an *H* atom and a water molecule without any chemical bonding between the electrode and the hydrogen adsorbate.

However, this theorization only reached the scientific society in the mid-twentieth century after Butler's correction of the activation energy barrier calculation, considering the attractive interaction between each particle [14].

Butler's paper was analogous to that of Gurney (in fact he predicted his colleague's fault) considering the *H* atom chemically bonded to the metal after the electron transfer by the proton in H_3O^+. Consequently, the overpotential projected by Butler was lower (ca. 1 eV) and the process became more plausible. Depending on the nature of the metal, the studied reaction ($H_3O^+_{(aq)} + M_{(Metal)} + e^-_{(Metal)} \rightarrow M[H]_{ads} + H_2O_{(l)}$) would occur with more probability for a metal exhibiting a tougher bonding with *H*. In fact, this is the basis of Electrocatalysis (Figure 5). However, in 1936, there was still no open credit for this phenomenon of metal-hydrogen bonding in Electrochemistry as Catalysis in Chemistry.

$$H_3^+O_{(aq)} + M_{(Metal)} + e^-_{(Metal)} \rightarrow M[H]_{ads} + H_2O_{(l)} \qquad [I]$$

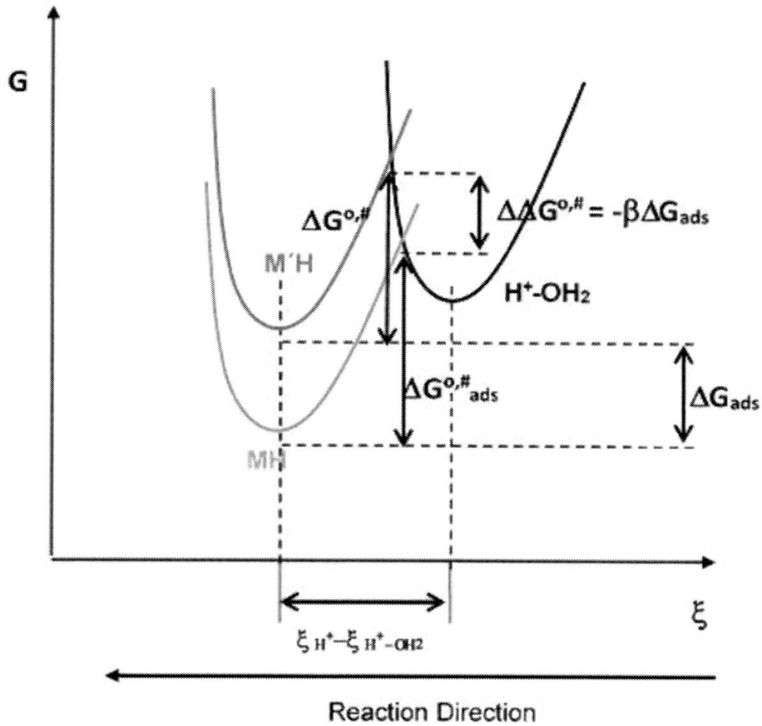

Figure 5. The free Gibbs energy, *G*, *vs*. position, *ξ*, diagram of the H^+-OH_2 electrocatalytic reaction: $H_3O^+_{(aq)} + e^-_{(Metal)} \rightarrow [MH]_{ads} + H_2O_{(l)}$.

Moreover, the relationship between the current density and the exponential term with the overpotential was firstly introduced by Agar in 1938 [15]. Nevertheless, ca. 23 years were necessary to use it as an electrocatalytic concept for fuel cells by Grubb working in the General Electric Company [16].

Figure 5 shows the potential energy vs. position diagram of reaction [I]. The two curves on the left depict the potential energy arcs for the *H* adatom interaction with two distinct metal surfaces. The height of the first intersection point with *M'* ($\Delta G^{o,\#}$ free energy of activation) of the *H* adatom/metal minus the lowest vibration energy level of the H^+-OH_2 bond is the height of the activation energy for the proton discharge reaction. As the strength between the H adatom and metal increases, *M* metal, the minimum of the curve diminishes with respect to the lowest vibrational level of the H^+-OH_2 bond. Then, the reaction has a lesser activation energy ($\Delta G^{o,\#}_{ads}$) and the reaction goes faster for the highest *H* adatom/metal interaction. The difference between both situations is the adsorption energy between the metals, *M* and *M'*; ΔG_{ads}. The most important consequence is that the change in the free energy of activation due to the adsorption process is proportional to the free energy of adsorption by the symmetry factor, β; $\Delta\Delta G^{o,\#} = -\beta\Delta G_{ads}$. The intersection point of the three curves describes the change in the free energy of activation, $\Delta\Delta G^{o,\#}$, that yields a faster reaction rate for an *H* adatom/metal bond formation and a stronger bond implies a larger $-\Delta G_{ads}$. The good adsorber with the highest hydrogen/metal bond produces a lower $\Delta G^{o,\#}$ and a faster reaction as predicted by Electrocatalysis. Graphically it was possible to demonstrate that $\Delta\Delta G^{o,\#}$ is a β fraction (and often ½ for elementary reactions) of $-\Delta G_{ads}$ (Figure 5).

However, Horiuti's report was more explicit in pointing out that the rate-determining step in the hydrogen evolution reaction is the proton discharge [17]. It was shown that the reaction rate on various electrodes would depend primarily on the strength of the bonding between the *H* adatom formed onto an empty site on the metal catalyst [17, 18].

In the case of *RedOx* solution species in an outer electron transfer reaction, the activation energy can be better described using the Marcus theory (Nobel Prize in Chemistry, 1992) [19]. His studies on unimolecular reactions using transition and collision theories committed him to elaborate in 1952, producing the Rice-Ramsperger-Kassel-Marcus (RRKM) theory [20, 21]. It is an extension of the previous RRK theory proposed between 1927 and 1928 [22, 23]. Moreover, the theory was further extended by Marcus for inner electron transfers [23-25]. The final theory, the *Marcus-Hush* theory, has been the widest and most accepted theory for Kinetics overviews in *RedOx* reactions with species in solution. The theory was presented by classical kinetics for electron transfer and the quantum nature of the process is almost shielded using other related concepts. This accepted formulation was not obscured by the theory on electron quantum mechanics developed by Jortner and Kuznetsov [26-28] for no radiant transitions. This was due perhaps to its limited application to heterogeneous reactions, where one phase electron transfer cannot be used.

Nevertheless, the theories have been developed considering only the vibrational effects of the electron transfer and particularly the vibronic Piepho-Krausz-Schatz (PKS) Hamiltonian element.

The most interesting feature of the kinetic constant expression is the quadratic formulation for the free energy of activation including λ, the reorganization of free energy, for either the inner and outer spheres, which in a Morse potential energy curve leads to the Marcus inversion region. The quadratic potential energy variation of the initial and final states is the result of the approach to the Morse curves by parabolas. The two quadratic dependences with the reaction coordinates are then, introduced into the Arrhenius expression between the activation energy and the rate constant.

In fact, the real situation in electrocatalytic equations is more complex since the adsorption of any species (or all) has to be taken into account (Figure 5). Thus, if both the reactant and product are adsorbed on the electrode surface, the free energy of stabilization is not λ but a change between the adsorbed reactant and the adsorbed product; $\Delta\Delta G_{ads}$ [29-31].

More properly the overall kinetic equation contains a term that can be associated with the charge transfer coefficient, α, of the entire process and the symmetry factor, β, in Figure 5. Using this coefficient, many possibilities are likely and also a partial charge transfer coefficient between adsorbed reactant and product can be foreseen. Then, lower Tafel slopes than 0.06 V decade^{-1} can be explained through this dual adsorbed state charge transfer with a Temkin isotherm state.

The changes in the activation energy due to the adsorption of reaction products were shown before [31, 32].When the product is adsorbed at the electrode surface, the activation energy $\Delta G^{o,\#}$ decreases in magnitude and there is a displacement in the relative position due to the ground state modification of configuration (Figure 5). This implication goes on the contrary when the reactant is adsorbed on the surface. The problem of both reactant and product attached to the catalyst defines the real change in the free energy of adsorption, $\Delta\Delta G_{ads}$. If we consider that the intersection curves are nearly straight, the contribution to the activation energy will be $\pm\beta\,\Delta\Delta G_{ads}$ with β constant over a wide range of $\Delta\Delta G_{ads}$.

The entropic factor is usually neglected in the concerns of activation energies. However in the case of electrocatalytic reactions with adsorbed compounds it is not possible at all. $T\Delta\Delta S_{ads}$ will significantly contribute to the activation energies due to the bond formations and ruptures as a consequence of adsorption. However, the thought that this magnitude can change similarly to the enthalpic component is rather inconsistent. The change in the entropy of adsorption would probably involve a gradual variation of rotations and vibrations together with bond formations with the electrode surface. In the case of statistical thermodynamics, the change in the partition functions along the reaction coordinate is usually denoted as multiples of the frequency factor, along which there is some change in the entropy due to adsorption. On the other hand, we can also say that once the bond to the electrode is formed is probable that relaxation energies will be expanded. Consequently, for strong adsorption cases, the $T\Delta\Delta S_{ads}$ term will go through a maximum of activation. These facts have been extensively discussed in [33].

According to the above assumptions, if both reactant and product are adsorbed on the electrode, the pathway will probably not be the rate-determining step. The strong dependence of ΔG_{ads} on θ suggests that is very important the special case of intermediate surface coverages. If we consider a linear variation between ΔG_{ads} with θ for both reactants and products under Frumkin conditions and when the interaction parameter, r is taken at intermediate values of θ, r is ca. 4:

$$\frac{\partial \ln\left(\frac{\theta}{1-\theta}\right)}{\partial \theta} = \pm r \qquad (56)$$

Thus, we can demonstrate that the Tafel slope between the electrode potential, E, and the logarithm of the current density, $\ln j$, will be:

$$\frac{\partial E}{\partial \ln j} = \pm \frac{RT}{(nF/\theta)+(n+\beta)F} \qquad (57)$$

These differential effects show that the change in ΔG_{ads} with E involves two factors; one only dependent on the electrostatic energy variation in $\Delta G^{o,\#}$ and the other indirectly dependent on E arising from the variation with θ. Since different factors can involve a certain Tafel slope, the usual common slopes of 0.06 or 0.12 V decade^{-1} are not accurate for an electrocatalytic reaction.

On the Electrocatalytic Behavior of a Metal Electrode Surface

One of the most popular plots to select an electrode catalyst and achieve a proper rate of an electrochemical reaction is the Volcano plot (Figure 6). It usually represents the logarithm of the exchange current density, logj_o, of the reaction (and in this case hydrogen evolution) as a function of the bond strength of *H* to the metal ($-\Delta G_{ads}$). The log j_o increases with an increase of *M-H* bond, $M[H]_{ads}$ [34].

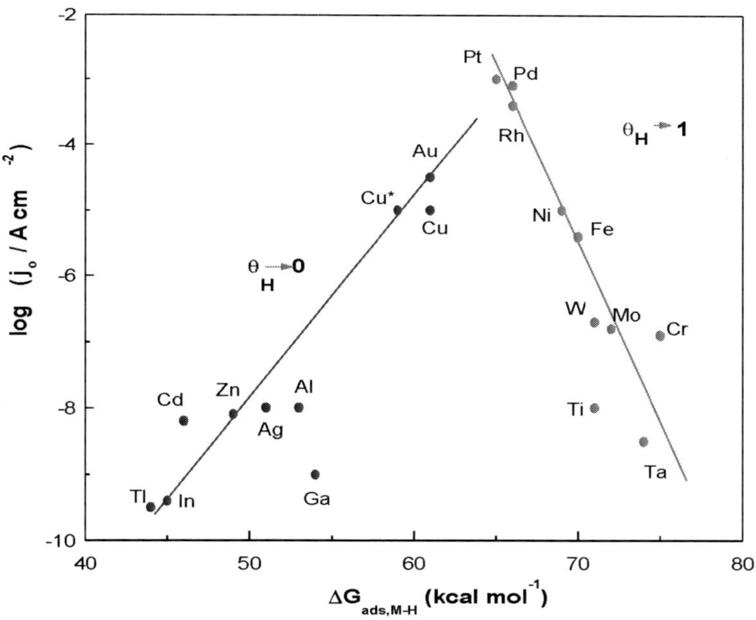

Figure 6. Volcano plot; log(j_o) for the acidic hydrogen evolution reaction as a function of the M-H bonding strength, $\Delta G_{ads,\,MH}$. Cu* denotes other crystalline orientation of Cu.

As the absolute value of ΔG_{ads} increases, the rate of the "pseudo-equilibrium" velocities at the reversible potential gets more rapidly. Of course, an increase of *M-H* bond strength means that the steady-state surface coverage (θ) by *H* on the metal *M* also increases. Thus, as the empty sites on the electrode fill up, the direct discharge of H^+ onto one of them will be replaced by the discharge of H^+ on top of an absorbed H site (high value of θ). This process yields molecular hydrogen in only one step (Heyrovsky path) [35]:

$$M[H]_{ads} + H^+_{(aq)} + e^-_{(Metal)} \rightarrow H_{2(g)} \qquad \text{[II]}$$

If Heyrovsky's reaction becomes the rate-determining step, the H filling tends to cover most of the surface ($\theta_H \rightarrow 1$). Therefore, it will now be more demanding for the hydrogen evolution reaction to occur on increasing amounts of M-H adsorbates. Then, the production of molecular hydrogen starts to slow down and a maximum in the rate of the reaction will occur as shown in Figure 6. As it is indicated, there are 2 types of metal catalysts, one with a positive slope with almost $\theta_H \rightarrow 0$ up to the 1B Group metals and the other from the largest j_o value with a negative slope of $\theta_H \rightarrow 1$ with the Platinum Group metals. Thus, platinum is the one for optimal conditions of rate and adequate *MH* bond strengths.

This plot is also completed using any current distinct to j_o, however, the first author that introduced this term in Electrocatalysis was Agar [15]. This kinetic parameter is the velocity either of the forward or the backward reactions under reversible conditions, that is, each of them in absolute value. Thus, the total current under irreversible conditions is the difference between the partial current density in forward $\vec{j}$ minus that of the $\overleftarrow{j}$ backward direction. According to Electrochemistry, it can be explained as the difference between oxidation and reduction reactions. The total current density outside equilibrium, j, is the difference them:

$$j = \vec{j} - \overleftarrow{j} = 0 \qquad (58)$$

Therefore, it is quite rational to say that j_o is the (equal and opposite) current density in each direction at true reversible conditions ($\eta = 0$). In this sense, j_o is a real finite number, e.g., 10^{-3} A cm^{-2} for the hydrogen evolution reaction on a polycrystalline smooth platinum electrode at 25°C.

With any current or reaction rate, it is a "Volcano relation" which compares it with respect to the adsorbate bonding on each substrate (as an intermediate). The term "bonding" also includes the heat of sublimation of the "catalyst electrode". Table 1 shows the melting points of a number of metals, M, with the most interesting platinum-type metals in the middle of the group. The adsorbate, M-X, (being X the species attached to M) will bond well enough, but not too much so that reactants just adhere to the surface and do not desorb.

Table 1. Melting point (MP) of some catalytic metals (°C)

Metal	MP (°C)	Metal	MP (°C)
Al	660	Pt	1770
Au	1063	Rh	1996
Cu	1084	Ru	2427
Ni	1453	Ir	2454
Pd	1550	Os	2700
Lu	1760	W	3380

Apart from platinum's "intermediate" nature on bonding, another point in platinum's favor is the availability, since platinum can be purchased in various suitable forms at a "reasonable" price. On the other hand, some other noble metals are difficult to find and/or purchase. The word "noble" means here *stable* and it is a good starting point for an electrocatalyst. It must be stable itself, thermally and electrochemically. On the last point, platinum is only fairly good because oxide-free platinum starts to dissolve around 1.0 V referred to the normal hydrogen reference scale. By using it in anodic reactions in a potential range higher than 1.0 V, Pt(II) is likely to solubilize into the solution and may be later deposited on the cathode. In fact, the formation of these hydroxyl and oxo-containing species is responsible for the catalytic behavior for many anodic oxidation reactions. This is due to the characteristic of being precursors of incipient surface oxides that promote oxidation. However, the inclusion of other metals as alloying agents, such as ruthenium, also induces oxidation since they are able to form hydroxyl species at lower potentials than pure platinum, being a better electrocatalyst than pure platinum itself.

Electrode Kinetics Far from Equilibrium

There are at least two kinetic parameters to evaluate, j_o and β in the case of an elemental reaction of solely 1 electron transfer (or α the electronic charge transfer coefficient in a multiple step reaction).

In Figure 7, the η *vs. log j* plots shows two electrocatalysts *A* and *B* compared with the same j_o and dissimilar β's. Their relative performance depends on the operating potential, and we cannot affirm which is the best unless the potential is previously stated.

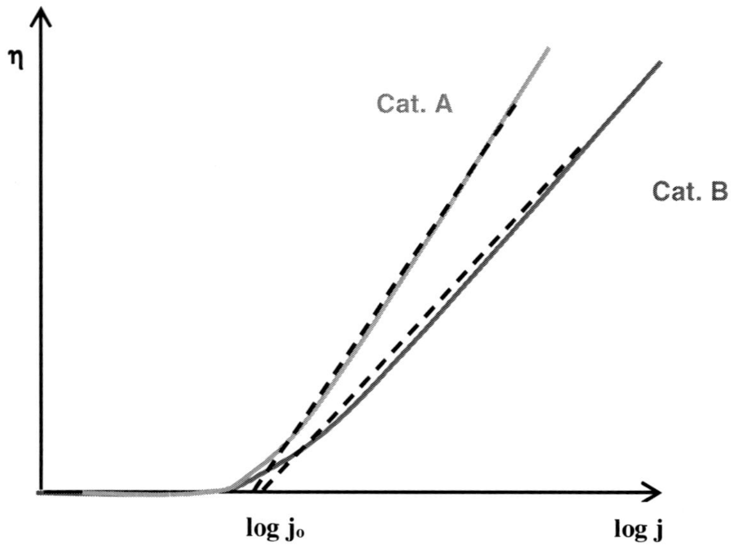

Figure 7. Tafel lines for a simple process comparing two electrocatalysts *A* and *B* with the same j_o and different β s.

It is important to mention that there is no confusion in the employ of either the symbol α in place of β. The latter indicates the electric contribution to a simple and unique reaction, and α to processes with multi-step reactions including this quantity β on the rate-determining step, the stoichiometric coefficient, the total number of electrons, and other quantities [20].

Instead, in Figure 8 we can see similar Tafel lines with the same β but distinct j_o. Thus, the ratio of j_o's would give the same result as comparing the relative current densities at any potential, including the reversible value at which j_o obeys.

Figure 8 indicates the methodology used to know which of the two electrodes, *A* or *B*, would exhibit the best performance for a given electrocatalytic reaction under certain experimental conditions. Since we need the highest currents at the working electrode potential (or overpotential η) it is clear that catalyst *B* will develop larger current densities than catalyst *A* at any selected η'. The difficulty of only considering the value of j_o as a diagnostic criterion (Figure 7) will yield confounding results at high current densities due to their different Tafel slopes.

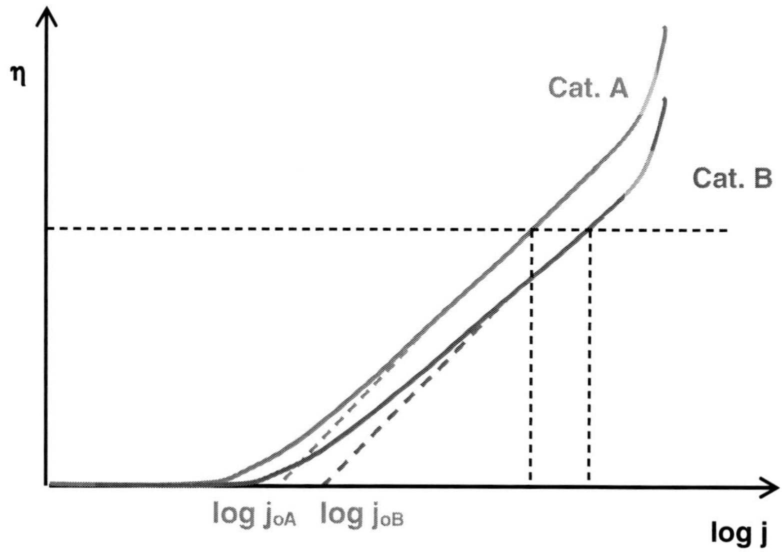

Figure 8. Tafel lines for a simple process comparing two electrocatalysts *A* and *B* with the same βs and distinct j_o.

The temperature is also an important factor and sometimes it happens that a desired reaction is only viable at low temperatures rather than at room values because of decompositions, formation of high oxidation state oxides, polymers, or a change in the mechanism of the reaction.

On the Mechanism of Electrocatalytic Reactions: The Rate Determining Step

In the analysis of Electrode Kinetics, there is usually one step that determines the rate of the whole reaction running at steady state. In this sense, the selection or design of a proper catalyst for the reaction simply involves the acceleration of this path. Since all former reactions need to go through this slow process, it is not possible to increase the rate, so this sluggish path yields an inhibition or blockage of the center site stablishing a rate-determining step (*rds*). Thus, either an increase in this rate or going through an alternative new path is what we call Catalysis. More precisely, it must be called Electrocatalysis since the path able to be modified is that one controlled by the electrode potential. Most of the molecules stay still until the blockage disappears or start going back to the first path to another alternative way to reach the products. Then, the molecules realize that there is no other path and they begin the reaction again. However, some of them are able to get to the final target, and other trip back in the sense of reaching the desired pace. Thus, the early reaction attains a "pseudo equilibrium" since the number of molecules waiting to get past the barrier becomes more or less constant. Finally, the entire pathway cannot get faster than this *rds*.

The analysis and control of this *rds* reaction is the electrocatalytic purpose of the study, e.g., by controlling the nature of the adsorbate, the nature and morphology of the surface and/or the electrode potential. In this sense, the first topic is analyzed through *in-situ* spectroscopic hybrid techniques (with simultaneous control of the potential) and the second by *ex-situ* or *in-situ* microscopic methods [36].

The *rds* has many aspects to consider and to be solved, that is, the nature of the reactant or product, the surface morphology at which it goes at an adequate rate, the proper adsorbed configuration and the electric potential energy to undergo. Thus, for the latter the nature of β is not possible to change for a given reaction path, only using a different nature of the electrode (metallic alloys or combined metals), however, the resulting electronic charge transfer coefficient, α, depends on other parameters. Moreover, there is one coefficient for the forward $\vec{\alpha}$ and another $\overleftarrow{\alpha}$ for the backward directions. Both as an addition gives the total number of electrons divided by the stoichiometric number. The selection of a certain electrocatalyst involves those coefficients, since obtaining large currents at a given electrode potential includes a high α (Tafel slope), either for the cathodic or anodic process.

Though, most of the simple and symmetric reactions (such as *RedOx* couples in solution) exhibit α ca. one-half. The choice of a certain overpotential to gain larger current density is directly connected to a higher α. However, the nature of the reaction sets the value of α on a certain metal. The subject of how to change α for a given electrode reaction mostly involves the type and morphology of the metal surface. Thus, the electronic population at the surface is different from the corresponding one in the bulk of the metal, and the crystallographic arrangement also indicates the mean electronic density on each metal atom, and then the possibilities of changing the activation energy. It is not quite simple to change the surface morphology of the electrode according to our necessities, but this is really *Electrocatalysis*. In fact, the easiest property to control is the atomic order and surface crystallography, since many good results have been obtained before [37, 38]. The problem is to maintain this "selected and proper" morphology at industrial levels as well. Thus, the development of roughness and porosity can also tend to disorderly surfaces not adequate to enhance the electrocatalytic reaction [39].

In a successive series of reactions, α increases as the consecutive partial reactions progress and also the current becomes larger since more electrons in the reaction are involved. Thus, the more complex (more intermediate pathways) the process is, the greater α we found, when being all electrochemical reactions. However, for pure chemical reactions a zero number of electrons are involved and no α coefficient develops, at least in the involved path. Only electrochemical pathways account for the α calculation along the mechanism.

On one hand, not only one reaction as *rds* has to be taken in the mechanism, since the optimization of the *rds* can modify the "adjacent" reactions. On the other hand, it is possible that after optimizing the surface morphology to accelerate the *rds* the rest of the reactions can develop other slow steps accordingly. Thus, it is possible to design different morphologies or crystallographic orientations on distinct surface sites to firstly keep the original morphology where the rest of the reaction paths occur at a sufficient rate, and secondly the new optimized morphology to accelerate the former *rds*.

The Electrocatalytic Process at Morphological Designed Metallic Surfaces

In the late 20th century, the employ of metal single crystals as electrodes gave new insights and solutions to the problem of accelerating an electrocatalytic *rds* reaction. Thus, we can find crystal (smooth or stepped) planes on which reactions occur faster than on others. Also, bimetallic surfaces with combined crystallographies can yield better results than pure single crystal metals. In the case of noble metals, a particular crystal face may provide a benefit of several orders of magnitude in rate constants compared with a polycrystalline material. This selectivity of a specific reaction renders positive effects for a certain metal catalyst after changing the crystal face depending on the reaction and also the electrolyte properties (Figure 9).

In the case of platinum, the pioneering work of Clavilier [40] after some improvements [41] gave rise to clean and repetitive cyclic voltammetric that can be used as fingerprints and as a selective index of what we need at a surface for a studied electrochemical reaction.The employ of a certain adsorption model for the configuration of H adatom and O adatom onto a platinum single crystal is important to Electrocatalysis. A langmuirian behavior is a simple relationship between a monolayer surface coverage (θ) to the bulk concentration in which we neglect heterogeneity and intermolecular force effects.

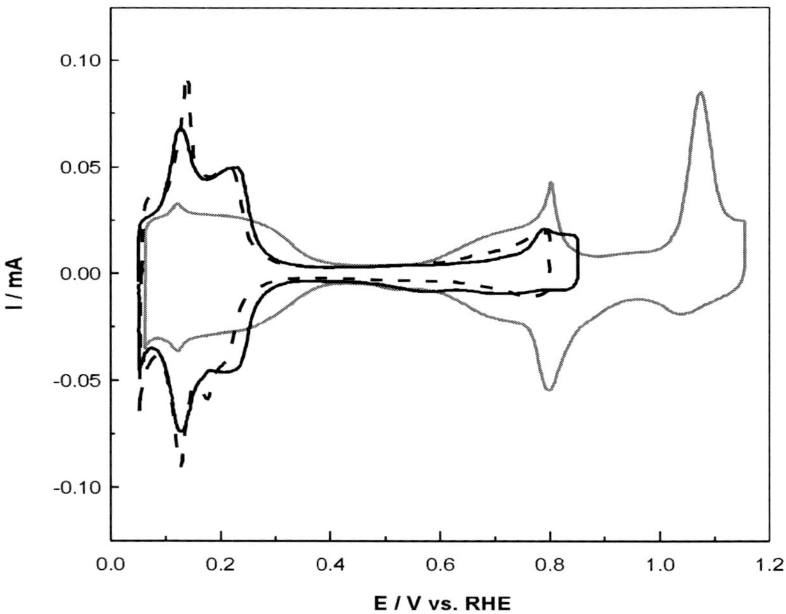

Figure 9. Cyclic voltammetric profiles of platinum single crystals, Pt(100) (cooled in air) black lines, Pt(110) dashed black lines and Pt(111) grey lines in 0.5 M sulfuric acid solution run at 0.05 V s^{-1} at room temperature.

To determine whether the process corresponds to Langmuir, Temkin, or a Frumkin isotherm is important for the mechanism analysis [42, 43]. However, at low and high θ, the isotherms give rise to almost the same behavior. If $0.1 < \theta < 0.8$, the reaction will request another action since the free energy of adsorption turns into a potential dependent process. Then, one should determine θ early in the investigation of a given system because the meanings of various mechanisms indicating factors change with the isotherm.

A recent review for predicting the low and high (stepped) Miller index surfaces of platinum based on the stable cyclic voltammograms in a supporting electrolyte is given in [44].

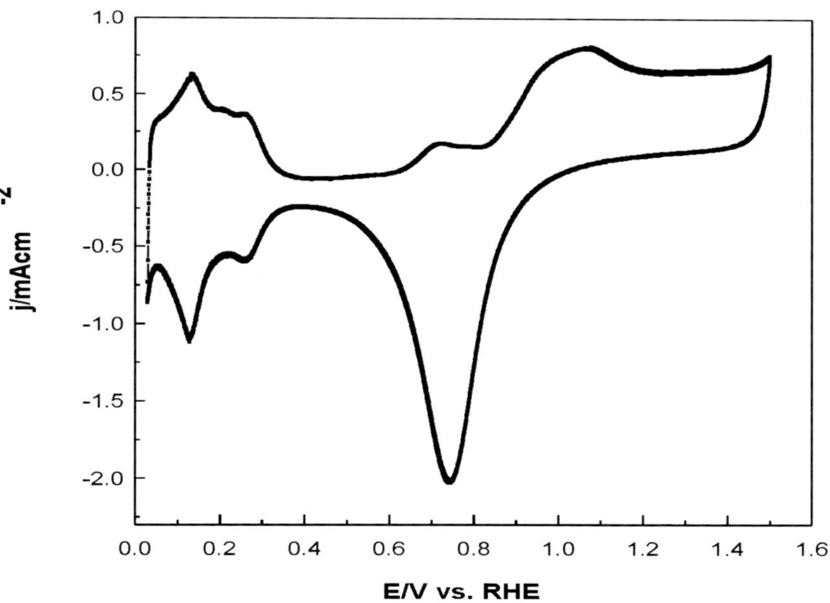

Figure 10. Stable cyclic voltammetry for columnar structured platinum was run between 0.02 V and 1.50 V at v = 0.010 Vs^{-1} in oxygen-free 0.5 M sulfuric acid at room temperature.

The preparation of single crystal faces is rather expensive and their stability upon large anodic potentials in industrial applications is quite doubtful. In fact, most of the surfaces are stepped planes that can only be prepared by laser techniques. Nevertheless, the preparation of large area ordered surfaces with a controlled roughness (factor from 10 to 10^4) and with certain crystalline orientations can be accomplished using electrochemical techniques (periodic potential programs) yielding stable electrodes for technical purposes [24]. Figure 10 shows the repetitive cyclic voltammetry profile of the platinum electrode after subjecting it to a square wave potential program with an upper limit of 1.7 V and a lower limit equal 0 V at 2 KHz in 0.5 M sulfuric acid. The electrode was characterized in [45] by several techniques and the calculation of the roughness factor was ca. 100. This surface increases its roughness with the application time of the square wave and also with the increase of the upper potential limit. The resulting electrode is characterized as a columnar periodic surface with a self-affine fractal behavior. The morphology can be mimicked using curvilinear revolution curves (as trochoids or cycloids) [45] i.e., points in a plane running along a single line direction without slipping.

Figure 11a shows the *ex-situ* scanning tunneling microscope (STM) image on the topographic mode of a columnar structured platinum surface. This quasi-cylindrical smooth surface was modelled by a trochoid as shown in Figure 11b.

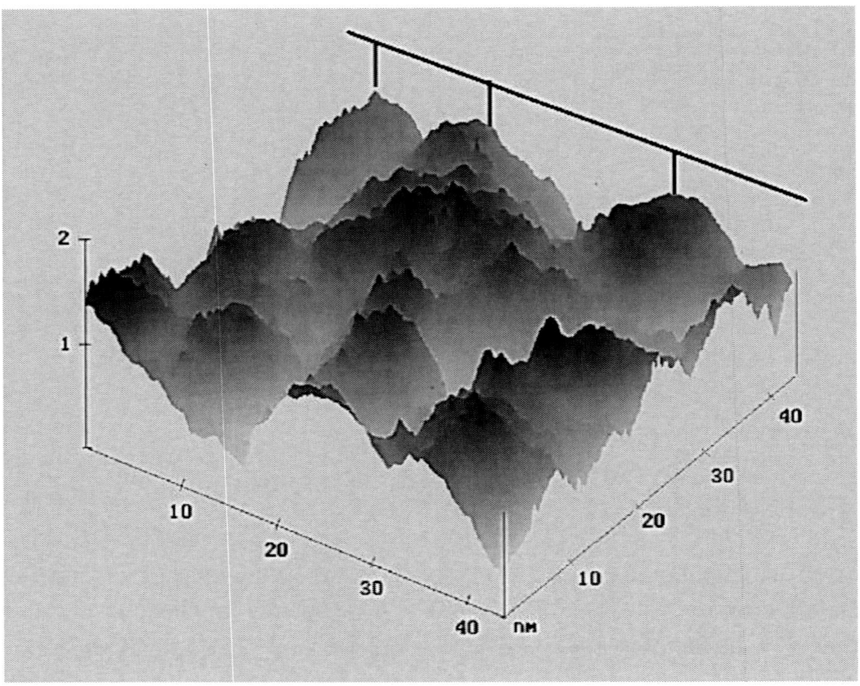

Figure 11a. *Ex situ* STM image in the topographic mode for a columnar structure platinum electrode.

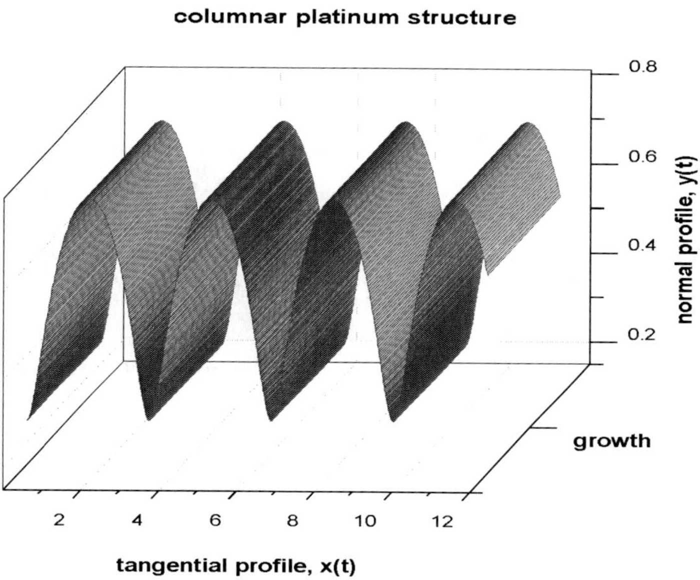

Figure 11b. Parametric representation of a periodic columnar surface (trochoid or shortened cycloid) with λ = 0.5, α = 0.5 (continuous lines).

The parametrization of the trochoid surface [45] is given by:

$$x(t) = \alpha(t - \lambda \sin t) \quad \text{and} \quad y(t) = \alpha(1 - \lambda \cos t) \tag{59a,b}$$

The mass and charge transfer equations involved in the electrochemical profile of the reactions occurring at these surfaces can be converted into a single parametric variable avoiding a 2D propagation of the current and potential functions.

When a surface is viewed under an electronic microscope it is possible to "see" that the apparently smooth material is not like that, i.e., irregularities and dislocations are part of the expected image. This is an important feature since the structure of electrode surfaces is a part of the interpretation in electrocatalytic reactions and can affect the amount of adsorbed species being the intermediates of the heterogeneous process. In this sense, all electrodes are in fact polycrystals, that is, each micro crystal has several crystallographic faces offering different electronic and bonding characteristics for intermediate species that complete a whole reaction on the surface. If one increases the degree of magnification, there is a kind of atomic roughness that can be seen [46]. Thus, two distinct levels appeared, one of a plane atomic dimension above another one, in a different direction forming a "*step*" rising slowly out of the surrounding micro flat surface.

Moreover, another form of irregularity on a surface that can be stochastically formed or intentionally produced is a *dislocation*. When an atomic row is occasionally affected by the absence of one of the atoms, the rest of the rows continue growing but the one that is missing causes the two rows surrounding it to close in and then fill the missing row with atoms. Such "dislocation" has effects for a while since the rest of the atomic rows are no longer straight. During this adjustment high local pressure develops near the dislocation [47]. Such sites tend to be "filled up" with adsorbates, but if their population is enough, they may be the origin of metallic breakdowns and fractures.

In summary, the "image" of a real surface is an ensemble of differently oriented crystals that justify the term polycrystal with atomic roughness and sometimes dislocations. Then, catalysis occurs by a different degree of adsorption energy for the dissimilar adsorbed atoms on the surface, weaker and stronger, and also with a different adsorbate configuration.

Surface diffusion also occurs between adsorbed atoms with high mobility forming new micro crystals with diverse shapes also converting into small ensembles along the surface. Thus, the formation of edges and kinks with large binding energies for the adsorbates can yield the catalytic effect of this restructured surface. Moreover, these processes are also affected by the local electric potential selecting a proper microcrystal promoting the formation of a defined adsorbed configuration with a given and certain catalytic behavior. This microstructure is the "atomic" heterogeneity characterized by a Temkin-type isotherm affecting the kinetics of the reaction since they are potential dependent ($d\theta/dE$). However, the surface heterogeneity yields distinct lateral interactions and the Frumkin condition also accounts. The complete isotherm results in a combined Frumkin-Temkin behavior with a single parameter of change in ΔG_{ads} with θ; ($d\Delta G_{ads}/d\theta$). The latter strongly affects the Tafel slope, but various mechanisms can exhibit the same slope, so a careful inspection is required. When applying an electrode potential different from the reversible value there are two contributions (forward and backward reactions) developing for the entire electrocatalytic process. However, the typical *0.059/n* slope is verified with small values of θ and a linear semilogarithmic relationship with the current ($dE/d\ln j$) is observed.

It is thinkable to derive various parameters for the change of the electrode potential with the bulk concentration, ($dE/d\ln C$), with θ ($dE/d\theta$), etc. that has to be encouraged to characterize well the reaction both for reactants and products. These parameters denote some keys to select a certain mechanism with respect to others. The *rds* often looks out from here and then, we need to describe its behavior with the definition of the reaction order at a certain electrode potential region. This order is obtain from the slope: $(d\ln j/d\ln C)_E$, and also with the type of isotherm and the nature and configuration of the adsorbed intermediate species. All these parameters need to be analyzed far from the mass transport region to avoid changes in the bulk activity and pressure. The determination of the *rds* and its good characterization in the middle electrode potential region, where Tafel's exponential law accounts is the basis of Electrodics. Thus, the kinetics and mechanism of the reaction considering the surface coverage dependence on the electrode potential is the real operative calculation of the Tafel slope.

In spite of those determinations, some criteria are supposed to be considered during the formation of the adsorbed species. In the example of the hydrogen evolution reaction, the first discharge Volmer's pathway (reaction [III]) involves the formation of the $M[H]_{ads}$.

$$H_3^+O_{(aq)} + M_{(Metal)} + e^-_{(Metal)} \rightarrow M[H]_{ads} + H_2O_{(l)} \qquad [III]$$

The finding of a good electrocatalyst involves the evaluation of the nature and energetics of the chemical bond between the adsorbate and the electrode, *M*. Moreover, the adsorbed water formation and the hydronium orientation in the double layer also play a key role in the energetics of the process [48, 49].

Nonetheless, due to its complexity this process can be taken as the rate-determining. Therefore, its stability must be intermediate since if the bond strength of the reactant to the metal catalyst *M* is very strong the reaction will stop there. In other words, the free energy of activation for the Volmer's reaction must be favorable so that the reaction is fast, but not so much to avoid anchorage of $M[H]_{ads}$. Therefore, it will depend also on ΔG_{ads}. The *H* atom needs to be recombined quickly together and so the intermediate θ at steady state will remain small.

However, with an increasing bond strength the rate constant of the Heyrovsky reaction will get faster, but the chemical recombination Tafel reaction will get slower. Thus, *H* atoms will be more difficult to be desorbed from *M*. At the same time with the increasing bond strength in *M*, the initial discharge reaction would not be rate-determining anymore, but the slowing of the rate constant in the second reaction, becoming *rds*. Anyway, the situation is solved by Heyrovsky's reaction where another hydrated *H* ion, H_3^+O, discharges directly on $M[H]_{ads}$ as [50];

$$H_3^+O_{(aq)} + M[H]_{ads} + e^-_{(Metal)} \rightarrow H_{2(g)} + H_2O_{(l)} + M_{(Metal)} \qquad [IV]$$

The recombination Tafel reaction between adjacent *H* atoms depends on the interdistance which varies with the nature of *M* and the θ value (which it also depends on the electrode potential). If this distance is too large, they are not able to recombine, but if it is too small, the ΔG_{ads} will not be favorable enough since their stabilization energies will repel each other in contact with *M*. Thus, the *rds* will be now this reaction but also dependent on the involved crystalline *M* plane, being strongly electrocatalytic in nature [51, 52].

If we mention Electrocatalysis, it means the finding of a new substrate with some changes in its surface (crystalline, nature, etc.), on which a specific reaction goes faster than the same reaction does on some standard surface. The most investigated electrode reaction is certainly the hydrogen evolution reaction ($2H^+ + 2e \rightarrow H_2$) and it has an exchange current density, j_o, of $10^{-2.3}$ A cm^{-2} on platinum and of $10^{-12.1}$ A cm^{-2} on mercury (Table 2).

Table 2. The logarithm of the exchange current density, log (j_o/A cm^{-2}) for the hydrogen evolution reaction on various catalytic metals

Metal	-log j_o	Metal	-log j_o
Hg	12.1	W	7.1
Pb	11.3	Fe	6.9
Ti	8.8	Cu	6.0
Sn	8.2	Ni	5.1
Cr	7.6	Pd	3.1
Mo	7.4	**Pt**	**2.3**

This is in acidic medium with low adsorbable anions, however, there is a strong dependence on pH as it has been shown for gold. Also, for the oxygen electroreduction reaction there is a change in the mechanism since the product of the reaction on gold is hydrogen peroxide in acid media, whereas hydroxyl complete reduction is found in alkaline electrolytes. Thus, the pH is more important than expected since it can change a complete mechanism or at least the *rds* of a reaction.

The robust electric field observed at the interface in electrocatalytic reactions is dependent on the nature and orientation of the water dipoles presence on the surface. If we talk about changing an electrode potential by 1 volt, it refers to a change of 1 volt delivered over 3-4 Å, i.e., one is dealing with electric fields of ~10^7 V cm^{-1}. These facts are intrinsic of the electrochemical interface and are sometimes not understood by other scientists. The extent of such fields distorts the silhouette of molecules adsorbed at the interface and evidently, a change in their reactivity or at least a modification of the enfolded and oriented molecules on the nearby is observed. Water in the double layer has a quadrupole characteristic and can change the local electric field at a certain "active site" within the overall electric field of the metal-solution interface.

The finding of an optimal electrocatalyst for a given electrochemical reaction is not the same as predicting its mechanism or its *rds*. However, if we know the mechanism and the reaction as being the rate-determining, it will be greatly helpful to determine the best catalyst. We have talked about single crystal surfaces and their defects intrinsically forming the complete surface (designed or not by us). There are certainly active sites that can conduct a defined path of the mechanism, while other distinct sites are more capable to conduct other paths, thus, the overall surface does not behave homogeneously. Then, specific planes on the single crystals have to be used to understand the process, which on first thought, the reaction is supposed to occur at a polycrystal or at a preferred oriented one.

Consistently, anions will be adsorbed on the various planes to different degrees and the amount and configurations will be also potential dependent. The latter implies that their local electric fields will attract or repel the presence of other ions on those planes. The best example is the case of gold since its double layer in acidic solutions permits the study with conventional techniques [53, 54]. Since this topic is rather outsized and beyond the scope of a single Chapter of this Book further discussions are going to be omitted.

The adsorbed anions reach the metallic sites from the double layer to the *IHP* (Inner Helmholz Plane) and their ability to escape from them to other active catalytic sites is determined by surface diffusion on the various faces. Obviously, all this occurs under high electric field strengths, which will change with the chosen potential at which the reaction is going to be run.

Moreover, the bonding of anions or the interested molecules or atoms to surfaces covered with adsorbed water can be also a *rds* in a way that inhibits the reaction rate. However, this is more dependent on the electric field developed at the interface, i.e., it has to be larger than 10^7 V cm^{-1}. In any case, this contribution has to be included in the Tafel equation to be considered, that is, in the Tafel slope, *b* or more properly in α. So, when α is high (various paths from the *rds*) the slope, *b*, at a given current density will develop a small overpotential (η = a + b log *j*) only when the ordinate at origin is small. The value of *a* is simply -*b* logj_o, but this means that (and since *b* is low) logj_o has a tiny contribution only when j_o is high (remember that the figure is always less than unity), e.g., logj_o for hydrogen acidic evolution is -2.3 on platinum and -12.1 on mercury.

The Employ of Hybrid Techniques to Envisage Electrocatalytic Mechanisms

The knowledge of the electrochemical reaction is much connected with electronic and interfacial properties that are simultaneously obtained using conventional techniques, such as cyclic voltammetry, impedance spectroscopy and transient methods. On the other hand, advanced forms of spectroscopy, specialized to observe surfaces in contact with water, have come into electrochemistry, increasing the level of understanding of the entire process. Moreover, much work has been conducted to approach UHV techniques to characterize the surface, before and after adsorption or reaction advances. However, the vacuum methods are not capable to describe the surface in contact with a double layer, as they are commonly employed in Catalysis and Physics. It was not until *Ellipsometry* was applied to Electrochemistry by Bockris and Reddy in 1964 [55] that it was possible to observe material adsorbed on an electrode surface whilst in contact with the solution.

The application of spectroscopic Ellipsometry in Electrocatalysis involves the determination of the state of the substrate in the presence of water in the double layer or an adsorbate by changes in the polarization state of the light incident at an interface. It is applicable for distinct electrode potentials and envisages the thickness of the thin film and determines the optical constants and refractive properties of the electrode in such conditions.

Some approaches by Ellipsometry demonstrate the existence of new active sites at platinum in the double layer region when the electrode surface is subjected to cathodization at -1 V or -2 V. This process leads to the formation of base centers with good catalytic behavior for the oxidation of small organic molecules [56].

Besides ellipsometry, two specific techniques made a revolution in the novel academic advances. The *on line* differential electrochemical mass spectrometry (*DEMS*) and the *in situ* Fourier transform infrared spectroscopy (*FTIRS*) have shown to be the most significant associations with conventional electrochemical methods in Electrocatalysis [57, 58]. The first one determines the nature of gaseous or volatile products formed during the electrocatalytic reaction, and the second one determines the nature and configuration of the adsorbed or desorbed species produced along the process.

DEMS by its *on line* mode is a technique that is able to detect the existence of atoms, molecules or their fragments (when volatile) by a quadrupole detector and identify in a mass spectrometer the quotient between its mass and charge while the potential is scanned at a relative medium sweep rate. The instantaneous analysis of the mass spectrometry data constitutes the direct link between the faradaic current and the reaction chemical nature of reactant and/or products [59, 60] for the proposed possibility of nature reaction species detection by Bruckenstein [61].

The *in situ FTIRS* mode offers the study and analysis of the electrocatalytic processes at a molecular level simultaneously to the application of a constant potential while the reaction is taking place. An important target was to solve the definition of the external reflections in the presence of an electrolyte onto a metal reflector surface. Bewick et al. [62] solved the task by using a thin electrolyte film on the electrode and a transparent window to the radiation, avoiding in this way the total radiant absorption by the solvent. However, this technique, named *EMIRS* (Electrochemically Modulated Infrared Spectroscopy) presented the inconvenience of modulation in the electrode potential to amplify the signals [63].

The introduction of the Fourier transform to the infrared spectrometers allows the elimination of the electrode potential modulation, since the noise was statistically suppressed by the accumulation of mean averaged spectra. In addition, the acquisition of the entire spectrum is simultaneous with the application of the electric potential, which contributes to make the measurement faster and more accurate. This *FTIRS* was firstly introduced by Pons [64, 65]

When considering the reflection at a perfectly polished surface, the electric field of the incident phase is a function of the polarization state of the light and the incidence angle. The dependence of the phase shift experienced by the reflected radiation on the polarization state of the light can be perpendicular to the incidence plane (s-polarized light) at ca. 180° for all incidence angles. It can also be a polarization parallel to the incidence plane (p-polarized light) where the phase shift depends on the incidence angle (varying from 60° to 90°). This fact makes the latter very sensible to the nature and configuration of adsorption states for the species attached to the electrode surface. Therefore, we can select and study separately the species adsorbed on the electrode (using p-polarised light) or in the bulk of the electrolyte (by s-polarised light).

Electrocatalytic Reactions with Technological Applications

The Electrocatalysis of Methanol to Carbon Dioxide

One of the substitutes of hydrogen as a fuel is methanol in the case of using fuel cells since the reversible normal potential, $E^{o}_{j=0}$ for the complete oxidation (-0.017 and -0.38 V at pH = 0 and 7) is close to that of the hydrogen system. However, it does not reach the large j_o ($\approx 10^{-3}$ A cm^{-2}) of hydrogen on platinum and complete oxidation is not always possible, even on platinum. On the contrary, the large number of electrons involved in the complete oxidation to carbon dioxide makes methanol a good election as fuel together with its liquid nature at room temperature. Since the amount of electricity that fuel cells produce is only dependent upon the presence of enough fuel, a fuel cell can take the place of a battery but with a much larger capacity for supplying electricity for more prolonged times . A battery has to carry their "fuel" with them on their metallic plates and it is fixed almost forever. Accordingly, when it is all used a recharge is required, but fuel cells take their fuel from separate tanks and oxygen from the air without any necessity of recharge.

An early investigation about the mechanism of methanol oxidation was conducted for the first time by Bagotskii and Vasiliev [66] from the famous group directed by Frumkin. They suggested that the now well-known Langmuir-Hinshelwood chemical surface path was the *rds,* occurring at an electrode surface between an $[OH]_{ads}$ with radicals arising from the dissociative adsorption of methanol.

Their mechanism exhibits the following *rds*:

$$Pt[C\text{-}OH]_{ads} + Pt[OH]_{ads} \rightarrow Pt[CO]_{ads} + H_2O_{(l)} + Pt_{(Metal)} \qquad [V]$$

where $[CO]_{ads}$ was proposed as an adsorbed intermediate. When this suggestion was first made, the *in situ* detection of adsorbed species by FTIRS was not available until the paper by Bockris et al. [67].

Moreover, further work [68-70] concluded that the surface oxidation yields on top bonded carbon monoxide at less positive potentials and bridge bonded carbon monoxide at more positive values.

The first surface degradation renders various dehydrogenated compounds that finally yields $[CO]_{ads}$:

$$CH_3OH_{(l)} + Pt_{(Metal)} \leftrightarrow Pt[CO]_{ads} + 4H^+_{(aq)} + 4e^-_{(Metal)} \qquad [VI]$$

Since it was found a bridge bonded formate species [71]:

$$Pt[CO]_{ads} + Pt[OH]_{ads} \rightarrow [COOH]_{ads}\text{-}Pt_{2(Metal)} \qquad [VII]$$

and,

$$[COOH]_{ads}\text{-}Pt_{2(Metal)} \rightarrow CO_{2(g)} + 2Pt_{(Metal)} + H^+_{(aq)} + e^-_{(Metal)} \qquad [VIII]$$

After some more work it was possible to detect single bonded formiate at larger potentials [72]:

$$Pt[CO]_{ads} + H_2O_{(l)} \rightarrow Pt\text{-}[COOH]_{ads} + H^+_{(aq)} + e^-_{(Metal)} \qquad [IX]$$

$$Pt\text{-}[COOH]_{ads} \leftrightarrow CO_{2(g)} + H^+_{(aq)} + e^-_{(Metal)} + Pt_{(Metal)} \qquad [X]$$

Those mechanisms showed that the Tafel lines have to delimit two slopes, as it is observed experimentally.

This mechanism of methanol electrooxidation avoids any mention of a "blocking" carbon monoxide as a poison, i.e., a carbon monoxide radical three-fold bonded to platinum with no more part in methanol oxidation to carbon dioxide. However, there is a feasibly building up of a carbon monoxide product or at least a decrease in the number of the active sites available for further acts of the reaction. The entire process then may occur in patches of the surface favorable to the fastest mechanism. Thus, it is important to study the problem on well-defined single crystal surfaces. For example, the reaction of bulk methanol oxidation is characterized by cyclic voltammetry on Pt(111) in sulfuric acid media (see Figure 12).

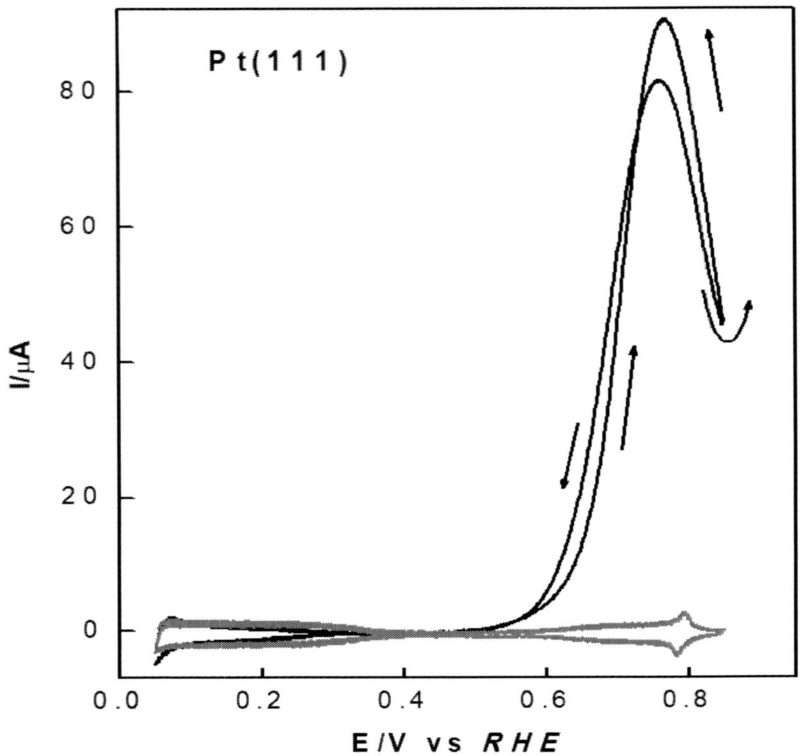

Figure 12. Forward and reverse bulk methanol electrooxidation on Pt(111) run at 0.02 V s^{-1} in 0.1 M methanol + 0.1 M sulfuric acid at room temperature. The voltammetric profile of the single crystal in the supporting electrolyte is superimposed as dark grey lines.

It is clearly seen that the potentiodynamic behavior of the process up to 0.85 V has a little hysteresis when run at 0.02 V s^{-1}, however, it depends on the nature of the supporting electrolyte, the scan rate, and the upper potential limit besides the crystalline orientation of the surface.

One of the first authors who discussed the problem of the influence of the electrode surface on the oxidation of organic molecules was Parsons [73].

He pointed out that the main oxidizing agent was $[OH]_{ads}$ coming from metal oxidation:

$$M_{(Metal)} + H_2O_{(l)} \leftrightarrow M[OH]_{ads} + H^+_{(aq)} + e^-_{(Metal)} \qquad [XI]$$

Thus, the electrochemical oxidation of methanol is strongly dependent on the existence of hydroxyl species and then, on the crystallographic properties of the studied surface [74, 75].

On the other hand, it was also found that in spite of being accepted the formation of adsorbed carbon monoxide through C atom, the other possibility of adsorption by O atom is also possible (Scheme 1). This Scheme shows the well-accepted Langmuir-Hinshelwood *rds* path, as well the methoxy formation with the final pathway as an Eley-Rideal reverse reaction found in other reactions [76]. The occurrence of each one of the mechanisms depends on the experimental conditions.

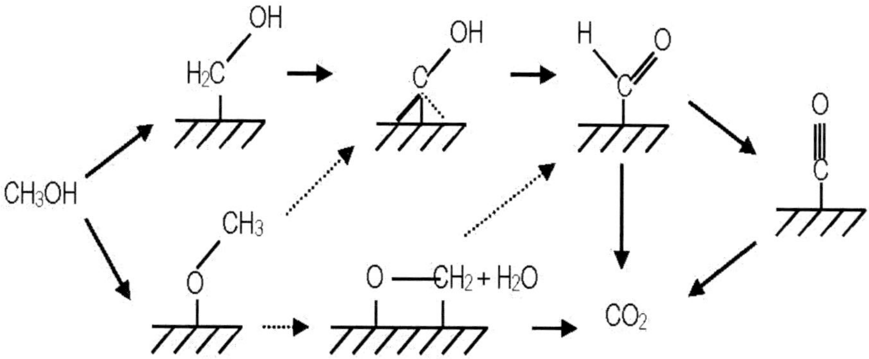

Scheme 1. General scheme of methanol electrooxidation considering series and parallel pathways yielding carbon dioxide as product. Solid and dashed arrow lines are the demonstrated and possible reaction pathways, respectively.

One of the most crucial experimental conditions to define the methanol oxidation mechanism is the effect of the crystalline orientation of the platinum electrode. Figure 13 shows the bulk electrooxidation of methanol on Pt(111) and Pt(110) in perchloric acid as supporting electrolyte.

Figure 13 shows the cyclic voltammetry at a relatively low scan rate with distinct upper limit potentials depending on the platinum single crystal stability. There is much to describe about the two panels, but one thing is for sure, the process at the (111) symmetry is inhibited after a certain potential value, i.e., 0.8 V, and it is dependent on the nature and concentration of the anion in the supporting electrolyte and of course, on the crystalline morphology of the surface. Thus, the inhibition is observed better in the reverse scan between 0.4 and 0.6 V, and at a lower sweep rate.

On the other hand, due to the presence of steps on the (110) symmetry the promotion of the reaction above 0.7 V is clear with a little inhibition up to 0.8 V. However, there is a "cleaning up" of the surface from adsorbed carbon monoxide between 0.7 and 0.3 V that produces a larger reverse current above the forward one. These facts made us think that there is a difference between the configuration of the adsorbed carbon monoxide on each surface or at least a distinct amount of them. Thus, this encouraged us to perform *in situ FTIRS* experiments to have a little more knowledge on the process.

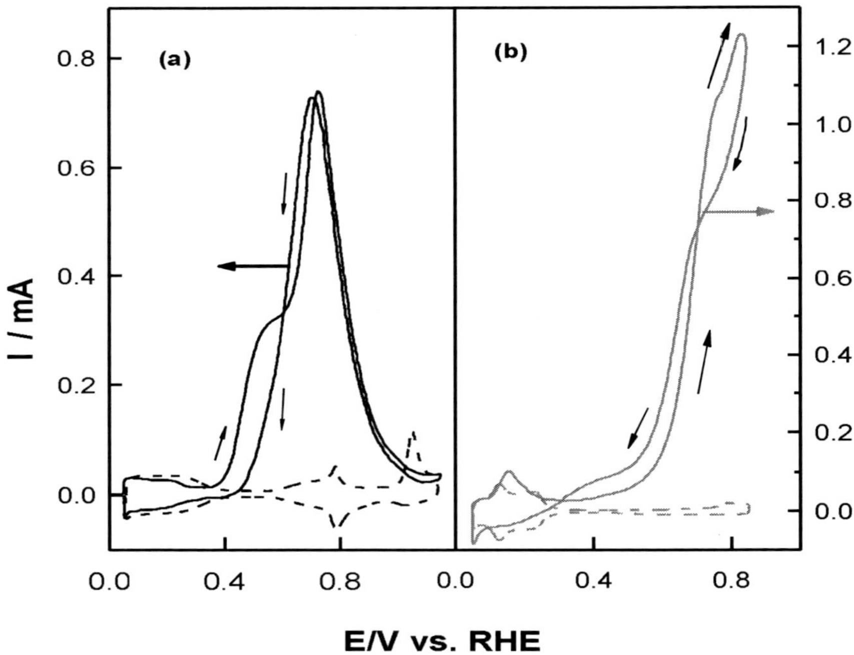

Figure 13. Anodic and cathodic scans for bulk methanol electrooxidation potentiodynamic curves on (a) Pt(111), black lines, and (b) Pt(110), grey lines, run at 0.02 V s^{-1} up to 1.15 and 0.82 V, respectively, in 0.1 M methanol + 0.1 M perchloric acid at room temperature. Black arrows indicate the scan directions. The cyclic voltammetric profile of each surface in 0.1 M perchloric acid is superimposed as dashed black and dashed grey lines.

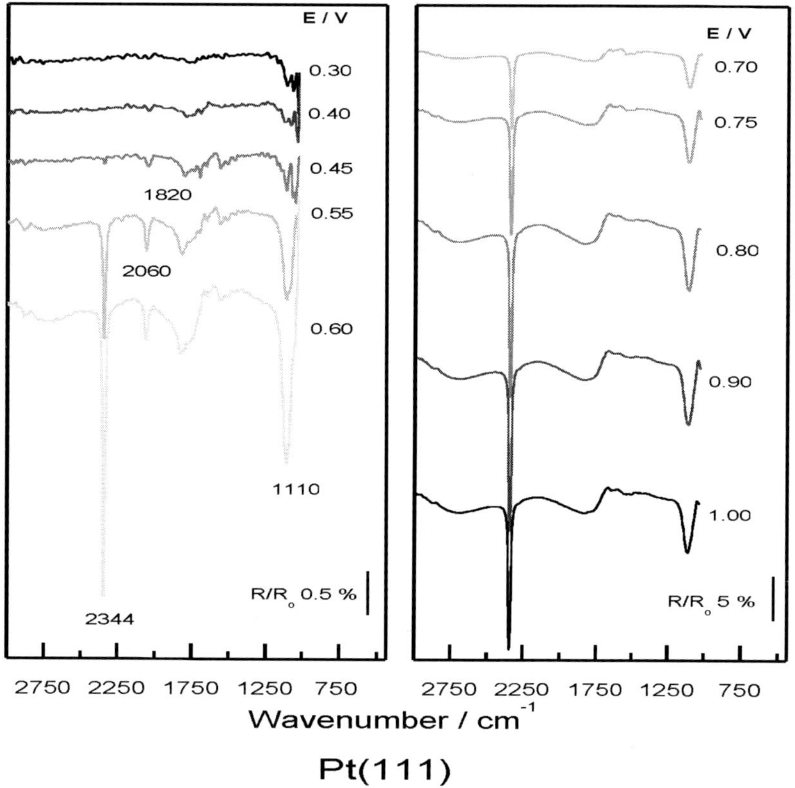

Figure 14. (Continued).

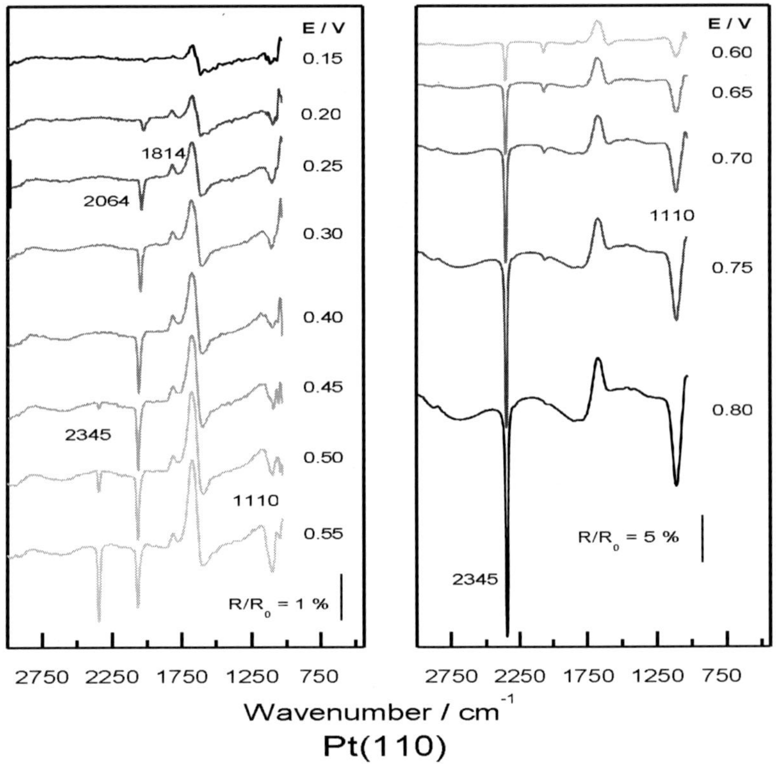

Figure 14. Increasing potentiostatic (step by step) *in situ* FTIR spectra in the 1150–2800 cm^{-1} range for bulk 0.1 M methanol + 0.1M $HClO_4$ for Pt(111) (upper panel) and Pt(110) (lower panel). Interferograms were obtained after 128 scans with 8 cm^{-1} resolution. The applied potentials and the characteristic bands are denoted in the plot. The reference spectrum was acquired at 0.05 V.

In Figure 14 we can see the bulk methanol oxidation products (consuming bands going down) in perchloric acid (symmetric stretching band at 1110 cm^{-1} of perchlorate). For Pt(111) at low electrode potentials (0.45 V) the bridge bonded carbon monoxide adsorbate is observed defined by the 1820 cm^{-1} band. The equivalent band on Pt(110) is observed 0.20 V before at 1814 cm^{-1}. This species is converted upon increasing surface coverage to the on-top bonded carbon monoxide vibrating at 2060 cm^{-1}. The complete oxidation product, carbon dioxide vibrating at 2344 cm^{-1} is observed above 0.55 V. On the other hand, on Pt(110) it appears at 0.45 V, in accordance with the potentiodynamic scan shown in the other figure. To avoid the interference of water and hydroxyl vibration signals, it is convenient to work with deuterated water as solvent to check for the equilibrium between both carbon monoxide adsorbed species. Moreover, in order to select the different signals of adsorbed or bulk solution species, it is convenient to employ p- and s-polarized light, respectively [77, 78].

The integrated band signals for FTIRS spectra as a function of potential are shown in Figure 15 (a) for the carbon monoxide adsorbed intermediate on both electrodes. The selection of bridge bonded adsorbate at potentials lower than 0.3 V and the bell-shaped contour for on-top adsorbate on Pt(110) (black lines) is denotable distinct to the tendencies on Pt(111) for both adsorbates (grey lines). However, there is a larger proportion of the amount of adsorbed species on Pt(110) than on Pt(111), but at lower potentials, reporting easier oxidative desorbing species on the first one (in spite of having larger amounts of carbon monoxide). The oxidative formation of carbon dioxide is observed in panel (b) depicting larger potentials necessary to form the species on Pt(110) than on Pt(111). Moreover, a bell-shaped integrated signal (grey lines) is shown for the latter electrode, which is expected for the poisoning of the surface by the strong adsorption of carbon monoxide between 0.60 and 0.90 V. The oxidation of adsorbed intermediates on Pt(110) is shown above 0.8 V (black lines) with an exponential growth.

As we have stated above the modification of platinum surfaces by foreign metal atoms promotes methanol oxidation either in UHV conditions or in the electrochemical environment. This new effect has been mostly discussed using the "third body model" [79], the "ligand effect" [80], or the "bifunctional effect" [81]. A theoretical review on the inclusion of metal reaction promoters was conducted by Anderson et al. [82] and later discussed in [83].

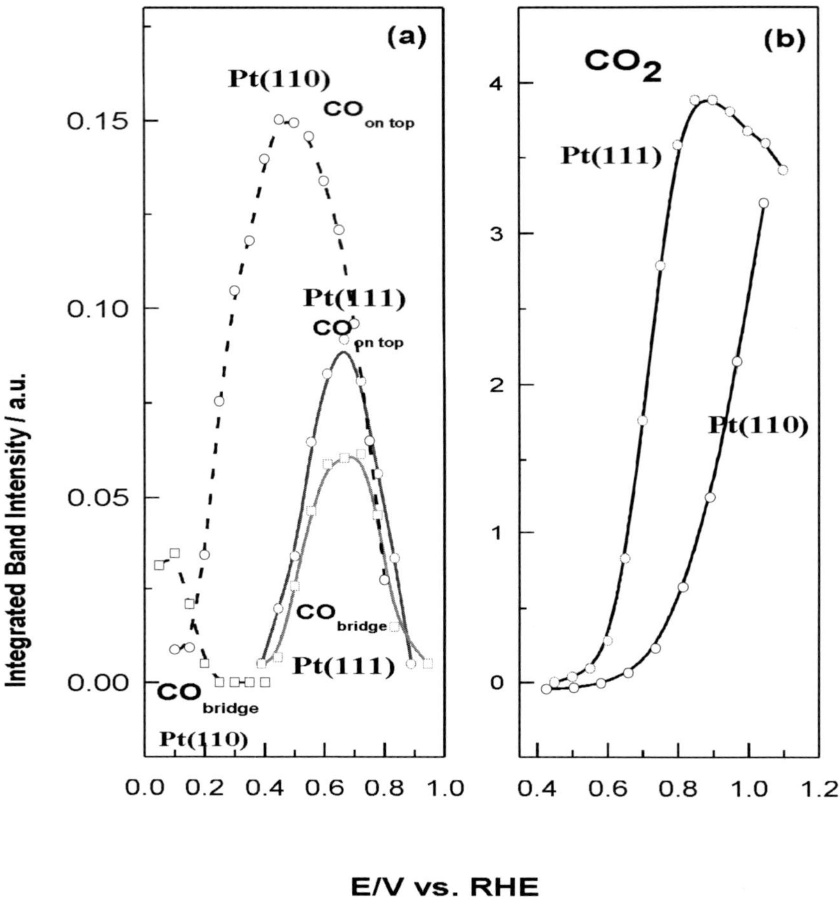

Figure 15. Integrated band infrared signals for (a) bridge and on-top bonded carbon monoxide and (b) carbon dioxide on Pt(111) (grey and dark grey lines) and Pt(110) (black continuous and black dashed lines) electrodes as a function of the electrode potential.

There was accordance for many years that in the case of platinum alloys, the two metals that are able to help methanol oxidation are ruthenium and tin. Methanol oxidation on Pt/Ru alloys, results in the production of carbon dioxide at lower potentials than those on pure platinum due to the formation of hydroxyl species at ca. 0.5 V necessary for the oxidation [84]. However, the presence of tin as Pt_3Sn alloys enhances methanol oxidation at low potentials by a faster first dehydrogenation and later carbon dioxide production detecting lower amounts of carbon monoxide [85]. The addition of tin (II) ions to previously adsorbed methanol produces a faster oxidation process than on the base surface as it was demonstrated by *DEMS* experiments [86].

In the presence of foreign metals, *M,* on platinum, the bifunctional mechanism proposes that *M* promotes the water discharge at lower potentials than pure platinum. Then again reaction [XI] applies.

These oxygenated species help the oxidation of carbon monoxide adsorbate, by either a direct chemical re-combination, reaction [V], or the following equivalent electrochemical pathway [41]:

$$Pt[CO]_{ads} + M[OH]_{ads} \leftrightarrow CO_{2(g)} + H^{+}_{(aq)} + e^{-}_{(Metal)} + Pt_{(Metal)} + M_{(Metal)} \qquad [V']$$

However, it is also possible the complete oxidation (Scheme 1) by the reverse Eley-Rideal mechanism when the intermediate is a methoxy species and no other metal is really needed [87];

$$Pt[OCH_3]_{ads} + H_2O_{(l)} \leftrightarrow CO_{2(g)} + 5H^{+}_{(aq)} + 5e^{-}_{(Metal)} + Pt_{(Metal)} \qquad [XII]$$

According to the ligand effect theory, the energy level of the modified substrate is changed to weaken the bond energy of the carbon monoxide adsorbate facilitating its oxidation. The best choice of the alloying or co-deposited elements depends on which step of the mechanism is the rate-determining step (normally reaction [V]). Therefore, also the molar fraction has to be 1:1 if this is true. Different results were obtained, so the discussion is still open.

If we simply consider the two main reaction pathways in the electrochemical Langmuir-Hinshelwood mechanism:

$$CH_3OH_{(aq)} + Pt_{(Metal)} \underset{k_{red,1}}{\overset{k_{ox,1}}{\leftrightarrow}} Pt[CO]_{ads} + 4\,H^{+}_{(aq)} + 4\,e^{-}_{(Metal)} \qquad [XIII]$$

$$Pt[CO]_{ads} + M[OH]_{ads} \xrightarrow{k_{ox,2}} CO_{2(g)} + H^{+}_{(aq)} + e^{-}_{(Metal)} + Pt_{(Metal)} + M_{(Metal)} \qquad [XIV]$$

We can check the kinetics through the following differential equation that yields the current density as a function of time, *I(t)*. This function is followed by potentiodynamic methods, so it is dependent on the surface coverage by the carbon monoxide adsorbate, $\theta_{CO,}$ being proportional to the maximum surface coverage that can attain the electrode surface, σ.

The balance for θ_{CO} yields:

$$I(t) = \sigma \frac{d\theta_{CO}}{dt} = Ba_{CH_3OH} k_{ox,1}(1-\theta_{CO}) - Ba^4_{H^+} k_{red,1}\theta_{CO} - A k_{ox,2}\theta^2_{CO} \tag{60}$$

B and *A* being kinetics constants:

$$\sigma \frac{d\theta_{CO}}{dt} + B\theta_{CO}\left(a_{CH_3OH} k_{ox,1} + a^4_{H^+} k_{red,1} \right) - Ak_{ox,2}\theta_{CO}{}^2 - Ba_{CH_3OH} k_{ox,1} = 0 \tag{61}$$

If the activities of methanol and proton are constants (those at the bulk of the electrolyte) we will obtain this condensed expression:

$$\sigma \frac{d\theta_{CO}}{dt} + \theta_{CO}\, K + \theta_{CO}{}^2\, L + M = 0 \tag{62}$$

with:

$$K \equiv B\left(a_{CH_3OH} k_{ox,1} + a^4_{H^+} k_{red,1}\right); L \equiv -A k_{ox,2}\,;$$
$$M \equiv -Ba_{CH_3OH} k_{ox,1} \tag{63a,b,c}$$

This expression has the form of a Riccati equation with electrode potential dependent coefficients:

$$\frac{d\theta_{CO}}{dt} + \theta_{CO}\frac{K}{\sigma} + \theta_{CO}{}^2\frac{L}{\sigma} + \frac{M}{\sigma} = 0 \tag{64}$$

Then, the solution is:

$$\theta_{CO} = \frac{\sigma C_1 r_1 \exp(r_1 t) + \sigma C_2 r_2 \exp(r_2 t)}{A k_{ox,2} C_1 \exp(r_1 t) + C_2 A k_{ox,2} \exp(r_2 t)} \quad (65a)$$

with:

$$r_{1,2} = \frac{1}{2} + \frac{K}{2\sigma} \pm \sqrt{\left(\frac{1}{2} + \frac{K}{2\sigma}\right)^2 - \frac{LM}{\sigma^2}} \quad (65b)$$

We can look at the values of r_1 and r_2 by taking the specific reaction constants:

$k_{ox,2}k_{ox,1} \approx 10^{-11}$, $(k_{ox,2})^2 \approx 10^{-12}$, $(k_{ox,1})^2 \approx 10^{-10}$, $(k_{red,1})^2 \approx 10^{-8}$, $k_{ox,1}k_{red,1} \approx 10^{-9}$

And then, we can obtain a simpler form:

$$r_{1,2} = \frac{1}{2} + \frac{B a_{H^+}^4 k_{red,1}}{2\sigma} \pm \frac{\sqrt{B^2 a_{H^+}^8 k^2{}_{red,1} + B^2 a_{CH_3OH} a_{H^+}^4 k_{ox,1} k_{red,1}}}{2\sigma} \quad (66)$$

By considering $(k_{red,1})^2 > k_{ox,1}\, k_{red,1}$, the roots are:

$$r_1 = \tfrac{1}{2} + \frac{B a_{H^+}^4 k_{red,1}}{\sigma} \text{ and } r_2 = \tfrac{1}{2} \quad (67)$$

And the values of θ_{CO} will be:

$$\theta_{CO} = \frac{\sigma C_1 \left(\tfrac{1}{2} + \frac{B a_{H^+}^4 k_{red,1}}{\sigma}\right) \exp\left(\tfrac{1}{2} + \frac{B a_{H^+}^4 k_{red,1}}{\sigma}\right) t + \sigma C_2 / 2 \exp(t/2)}{A k_{ox,2} C_1 \exp\left(\tfrac{1}{2} + \frac{B a_{H^+}^4 k_{red,1}}{\sigma}\right) t + C_2 A k_{ox,2} \exp(t/2)} \quad (68)$$

But r_1 is approximately equal to r_2 and to ½, then:

$$\theta_{CO} \cong \frac{\sigma}{2Ak_{ox,2}} \tag{69}$$

Thus, under this simplification, it is likely to envisage the value of the specific rate constant of the *rds* knowing the θ_{CO} as a function of potential (or time which is equivalent since the technique is potentiodynamic with a constant sweep rate).

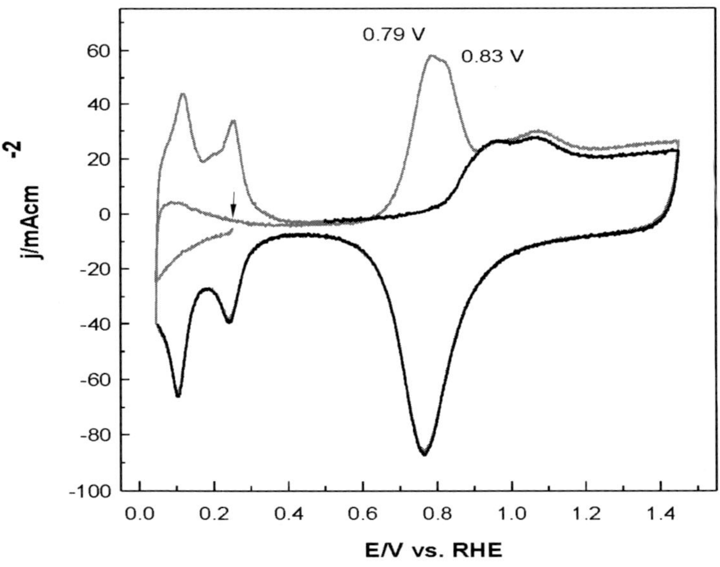

Figure 16. Adsorbed methanol residue electrooxidation (grey lines) by a potentiodynamic scan on polycrystalline platinum run at 0.05 V s^{-1} in 0.5 M sulfuric acid at room temperature. The voltammetric profile of platinum in the supporting electrolyte is superimposed as black lines. The adsorption of the residue was performed at 0.25 V for 5 min from a carbon monoxide-saturated acid solution.

The calculation of θ_{CO}, the steady state surface coverage by carbon monoxide residue, can be followed by anodic *stripping* voltammetry with the details given elsewhere [78, 84, 88]. By this technique, it is possible to separate the adsorption residue contribution from that of the bulk electrooxidation process.

The micro flux cell is adapted with a big flask containing the supporting electrolyte, which is used to wash the cell until there is no trace of methanol or carbon monoxide in the solution. The potentiodynamic current *vs.* potential profile is then run from the adsorption potential, firstly downwards to envisage the hydrogen electrosorption region and then, upwards to obtain the oxidation profile. The latter is called the *stripping* contour of the adsorbed residue. An example is presented in Figure 16 for methanol residue oxidation *stripping* on a large area of polycrystalline platinum surface using the potential of zero charge, 0.25 V, as the adsorption potential for methanol adsorption and oxidation.

It is clear that there are two adsorbed configuration contributions at 0.79 and 0.83 V assigned (using *in situ FTIRS*) to bridge and on top-bonded species of carbon monoxide, respectively. These results vary with the nature and concentration of the supporting electrolyte (perchlorate, sulfate, etc.) and of course by the presence of strong adsorbable anions, the crystalline orientation of platinum (low Miller indices, stepped planes, etc.) and also its roughness. Thus, the following anodic *stripping* experiment for methanol adsorption is conducted on a relatively large area platinum electrode (produced by electrochemical high frequency periodic treatments). The bright smooth surface of platinum can be increased with the application of fast periodic routines of square, sinusoidal or triangular waves at frequencies larger than 1 KHz [38, 89, 90]. This phenomenon is characterized by the platinum dissolution and redeposition to other center sites yielding distinct crystalline stepped planes depending on the upper and lower potential limits.

The final surface is stable upon extreme anodic treatments such as the oxidation of large aromatic molecules or polymers, and it is possible to use them in the industry without any problem. The cyclic voltammetry of the preferentially oriented electrode is similar to those of single crystals after incursions in the platinum oxide regions [91, 92].

The presence of carbon monoxide as an intermediate adsorbed species on platinum of a relative medium roughness behaves differently since there are certain stepped planes that shielded some adsorption configurations. Figure 16 shows the anodic *stripping* profile for methanol adsorption and oxidation starting the adsorption at the potential of zero charge (where the neutral species adsorbs preferentially). The *stripping* profile is wider and the anodic peak potentials for both single and 2-fold coordinated species are more closely.

The Elimination and/or Conversion of Carbon Dioxide to Fuels

One of the greatest hazards of our vulnerable world is the increasing carbon dioxide and methane (low hydrocarbons) atmospheric concentration. They are expected to rise to 500 ppm if we continue to burn fossil fuels or emanate its products from conventional industries. During the last 10 years, it was observed an increase in melting glaciers; a rise in the ambient mean temperature with the vast toxic gas concentrations that will make the Earth's surface practically uninhabitable for human beings.

There are various methods to convert those incomplete combustion products to other not too toxic or pollutive. In a matter to reduce the extension of the discussion we can say that, on one hand, the fixing of atmospheric carbon dioxide by converting it to a simple hydrogen carrier (liquid methanol) is one of the most promising ways. On the other hand, its conversion to methanol by adsorption and electroreduction is the other one but with some difficulties.

In this sense, we have made some contributions to use the same electrocatalysts (platinum and their alloys) for the reconversion of carbon dioxide to methanol and methylformiate as those for the fuel cell conversion of energy [93- 95]. The new finding is based on the fact that carbon dioxide adsorption in the hydrogen voltammetric region of platinum produced reduced species (Figure 16) that can be further converted into methanol. The nature of the reduced carbon dioxide and its electrooxidation on platinum has been followed by cyclic voltammetry and *in situ FTIRS* on distinct crystalline surfaces [93]. Three dissimilar "reduced" adsorbates were found in acidic solutions involving different extents of adsorbed units with weakly and strongly bound hydrogen atoms. The strong adsorbable intermediate, chemically similar to carbon monoxide and formyl species, was formed with difficulties to convert to methane or methanol.

The potentiodynamic scan for the oxidation of reduced carbon dioxide adsorbates formed on platinum in the $0.05 \leq E_{ads} \leq 0.35$ V potential range for $30 \leq t_{ads} \leq 600$ s exhibits a double current peak at 0.82 V and a shoulder at 0.73 V (Figure 17 (a)). At potentials higher than 0.95 V, the anodic scan resembles that of the blank in the supporting electrolyte, hence at this potential, the complete desorption of the reduced carbon dioxide species is attained. The amount of the adsorbed species at the surface was calculated from the electrooxidation charge density and it resembles that of a langmuirian isotherm (Figure 17 (b)). However, for $E_{ads} \leq 0.15$ V, a maximum and constant charge density equal to 194 μC cm^{-2} is found. For E_{ads} > 0.15 V, a firm reduction of the amount of adsorbate is detected. Similarly, the influence of t_{ads} was studied at E_{ads} = 0.10 V in Figure 17 (c) and for t_{ads} = 300 s the maximum coverage is attained [94]. Similar results were found for rhodium (Figure 18) but with more negative E_{ads} where carbon dioxide was dissociatively adsorbed as strongly bound carbon monoxide. The latter is subsequently hydrogenated resulting in *[COH]*$_{ads}$ and finally desorbing as carbon dioxide upon more anodic potentials.

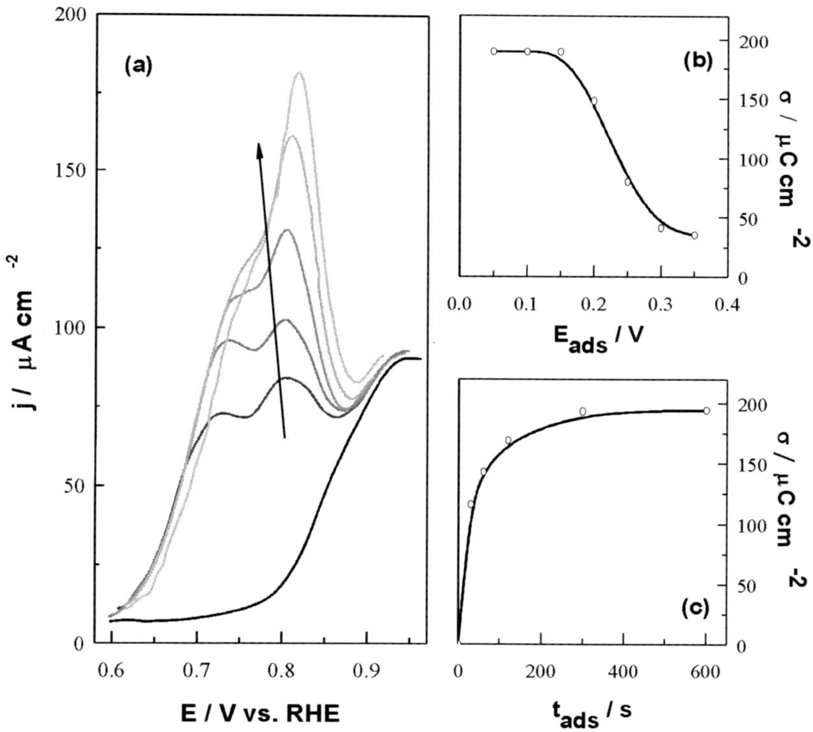

Figure 17. (a) Anodic potentiodynamic scan, *j vs. E*, of carbon dioxide reduced species on a polycrystalline platinum surface in 1 M aqueous H_2SO_4 saturatedwith CO_2 after adsorption at E_{ads} = 0.10 V for $30 \leq t_{ads} \leq 600$ s (arrow indicates the increasing t_{ads}). The blank is depicted as black lines. The sweep rate was 0.10 V s^{-1}; (b) charge density, σ *vs.* E_{ads} plot for t_{ads} = 300 s; (c) saturation values of σ *vs.* t_{ads} plot at E_{ads} = 0.10 V.

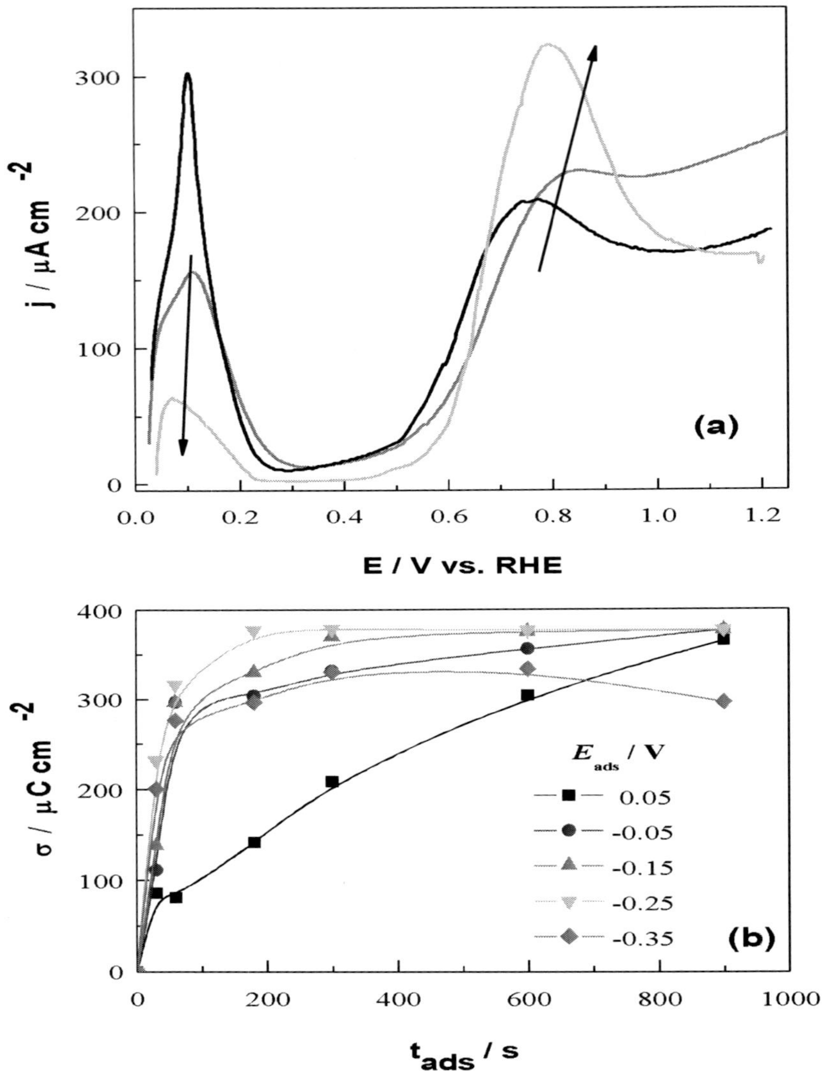

Figure 18. (a) Anodic potentiodynamic scan, *j vs. E*, of carbon dioxide reduced species on a polycrystalline rhodium surface in 1 M aqueous H_2SO_4 saturated with CO_2 after adsorption at E_{ads} = -0.15 V (pale grey lines) and 0.05 V (dark grey lines) for t_{ads} = 600 s. The blank is depicted as black lines and the arrow indicates more negative E_{ads} contours. The sweep rate was 0.10 V s^{-1}; (b) charge density, σ *vs.* t_{ads} plot for $-0.35 \leq E_{ads} \leq 0.05$ V.

In this sense, platinum is taken as an inactive catalyst for the reconversion of carbon dioxide to hydrocarbons or simple alcohols, so other transition metals of the 1 B family have to be employed. However, on carbon supported platinum and at low partial pressures, carbon dioxide is selectively reduced to methane at potentials near the reversible value (0.17 V) [95]. The problem is the low Faradaic efficiency (below 6% at the reversible potential), so a cathodic overpotential has to be applied. The proposed mechanism involves the surface reduction of carbon-containing intermediates by the on-top hydrogen adatom on platinum, being the onset of carbon monoxide co-adsorption with hydrogen near the reversible potential of methane.

The geometries of the adsorbed ensembles and binding energies of carbon dioxide on Pt(100) and Pt(111) surfaces were also calculated by semiempirical methods [96]. According to experimental data, the electric potential was mimicked by moving the Fermi level of the cluster surface together with the co-adsorption of hydrogen atoms for modeling the phenomenon. At sufficient negative potentials, it was observed the formation of side-on formate species with two carbon atoms bonded to the surface, regardless of the surface topology and the presence of co-adsorbed hydrogen atoms. The largest binding energies were found on Pt(100).

On the other hand, employing *in situ surface enhanced infrared absorption spectroscopy* (*SEIRAS*) and *on line DEMS* measurements on platinum electrodes, it has been reported that in the absence or presence of gaseous hydrogen, carbon dioxide reduction at potentials in the hydrogen electrosorption region results in linear and multiple bound carbon monoxide as the only stable adsorbate [97]. Thus, the formation and the following reduction reaction saturate the electrode at a surface coverage of 0.6, which is equivalent to about 0.45 ML of the total carbon monoxide coverage. The coverage and chemical nature of the carbon monoxide adsorbate, as indicated by the C–O stretching frequency, and their reduction current are largely independent of the presence of hydrogen, indicating that carbon dioxide reduction to carbon monoxide involves the reaction with atomic surface hydrogen rather than a reaction with dissolved gaseous hydrogen.

Concerning appropriate cathodes for a massive electrocatalytic reduction of carbon dioxide to methanol, indium and tin have good projections because each has a high overpotential for hydrogen evolution that is the parallel (and not desirable) reaction to the goal:

$$CO_{2(g)} + 6H^{+}_{(aq)} + 6e^{-}_{(Metal)} \rightarrow CH_3OH_{(l)} + H_2O_{(l)} \qquad [XV]$$

However, Hori [98] found the electrocatalytic reduction of carbon dioxide to methane on a molybdenum cathode to be also electrocatalytic. Therefore, a combination of a settled metal for coadsorption with hydrogen (such as platinum) and other metals (such as gold, copper, tin, etc.) for reduction will be necessary to design.

The visionary paper by Bockris [99] suggested the elimination of carbon dioxide emitting fuels in favor of hydrogen ("The Hydrogen Economy") to preserve the environment. May be this improve in the reconversion by electrocatalysis with platinum and their alloys with copper or gold complete the scheme.

There is tiny doubt that our outmost warning is the *Greenhouse Effect,* by then, the rise in temperature and danger to inhabitability would be in years and not in decades. Thus, the total substitution of carbon-containing fuels in favor of pure green or white hydrogen will be the main way to attain this goal.

Conclusions

The electrocatalytic reactions were described in terms of the existence of an adsorbed species, which was characterized by changes in the activation free energy and the adsorption free energy of the adsorbate. The current density *vs.* overpotential relationship was introduced by Quantum Electrochemistry using a variable Eckart potential barrier, from which the electron tunneling yields the empirical Tafel's law. Also, Gurney's Theory for the electron transfer at the electrochemical interface and Marcus's Theory for the *RedOx* reactions using the reorganization energy were described.

At least two kinetic parameters were involved: the exchange current density and the electronic charge transfer coefficient. The rate-determining reaction step for the hydrogen evolution reaction (Volmer-Tafel-Heyrovsky mechanism) was taken as an example with the variation of the Tafel slope with the type of isotherm (Langmuir, Temkin, etc.) for adsorbed hydrogen atoms. The most popular plot for selecting the most applicable electrocatalytic metal, the volcano plot, was described accordingly. The variations with the crystallographic orientation and the surface roughness were analyzed for platinum using cyclic voltammetry as a fingerprint of the surface.

On the other hand, the characterization of electrocatalytic reactions by hybrid techniques, such as ellipsometry, *on line* differential electrochemical mass spectrometry and *in situ* Fourier transform spectroscopy were employed for one of the most attractive reactions; methanol adsorption and oxidation on platinum.

The selected difference on the methanol and carbon monoxide behaviors at platinum single crystal surfaces were defined as typical electrocatalytic reactions (Langmuir-Hinshelwood and Eley-Rideal reverse paths). The anodic *stripping* experiments were used to determine the time-dependent surface coverage by the organic adsorbed residue. The alloy formation on platinum with other metals (able to adsorb preferentially oxygenated species) was taken as an electrocatalytic change of the pure electrode for the reaction enhancement. Finally, the possibility of carbon dioxide attaching and its reconversion to methane and methanol were described on distinct electrode materials.

Acknowledgments

The author wishes to thank his "old fashioned" professors who have helped him throughout his academic career.

References

[1] Tafel, J. (1905). *Z. Physikal. Chem.*, *50*, 641.
[2] Ashcroft, N. W., and Mermin, N. D. (1976). *Solid State Physics*, Philadelphia: Saunders College Publishing.
[3] Garoni, T. M., Frankel, N. E., and Glasser, M. L. (2001). *J. Math. Phys.*, *42*, 1860.
[4] Jang, N. D., and Kohn, W. (1971). *Phys. Rev.*, *3*, 1215.
[5] Odeh, F., and Keller, J. B. (1964). *J. Math. Phys.*, *5*, 1499.
[6] Wannier, G. H. (1960). *Phys. Rev.*, *117*, *432*.
[7] Bockris, J. O'M., Sen, R. K., and Mittal, K. L. (1972). *J. Res. Inst. Catalysis Hokkaido Univ.*, 20, 153.
[8] Eckart, C. (1930). *Phys. Rev.*, *35*, 1303.
[9] Eisberg, R. M. (1967). *Fundamentals of Modern Physics*, J. Wiley & Sons, New York.
[10] Bockris, J. O'M., and Khan, S. U. M. (1979). *Quantum Electrochemistry*, Plenum Press, New York, London.
[11] Slater, L. J. (1966). *Generalized Hypergeometric Functions*, Cambridge University Press.
[12] Artin, E. (2015). *The Gamma Function*, Dover Publications on Mathematics Inc., translated by Butler M, reprint edition.
[13] Gurney, R. W. (1931). *Proc. Roy. C., London, A134*, 127.
[14] Butler, J. A. V. (1936). *Proc. Roy. Soc.*, *A 125*, 423.

[15] Agar, J. N., and Dainton, F. S. (1947). *Disc Faraday Soc*, *2*, 170.
[16] Grubb, T. (1959). *U.S. Patent 2*, *913*, 511.
[17] Horiuti, J., and Polanyi, M. (1935). *Acta Physicochim U.R.S.S.*, *2*, 505.
[18] Eyring, H., Glasstone, S., and Laidler, K. J. (1939). *J. Chem.Phys.*, *7*, 1053.
[19] Marcus, R. A. (1982). *Faraday Discuss. Chem. Soc.*, *74*, 7.
[20] Marcus, R. A. (1952). *J. Chem. Phys.*, *20*, 359.
[21] Marcus, R. A., and Rice, O. K. (1951). *J. Phys. Colloid Chem.*, *55*, 894.
[22] Rice, O. K., and Ramsperger, H. C. (1927). *J. Am. Chem. Soc.*, *49*, 1617.
[23] Kassel, L. S. (1928). *J. Phys. Chem.*, *32*, 1065.
[24] Lin, S. H., Lau, K. H., Richardson, W., Volk, L., and Eyring, H. (1972). *Proc. Natl. Acad. Sci. U S A.*, *69*, 2778.
[25] Flowerscan, M. C. (1978). *J. Chem.*, *56*, 29.
[26] Wardlaw, D. M., and Marcus, R. A. (1985). *J. Chem. Phys.*, *83*, 3462.
[27] Kuznetsov, A. M., and Ulstrup, J. (1982). *Faraday Discuss. Chem. Soc.*, *74*, 31.
[28] Kuznetsov, A. M. (1982). *Faraday Discuss. Chem. Soc.*, *74*, 49.
[29] Bixon, M., and Jortner, J. (1982). *Faraday Discuss. Chem. Soc.*, *74*, 17.
[30] Gnanamuthu, D. S., and Petrocelli, J. V. (1967). *J. Electrochem. Soc.*, *114*, 1036.
[31] Plieth, W. J., and Vetter, K. (1960). *Z. Phys. Chem.*, NF, *61*, 282.
[32] Zinola, C. F., Castro Luna, A. M., and Arvia, A. J. (1994). *Electrochim Acta*, *39*, 1951.
[33] Appleby, A. J. (1985). "Electrocatalysis" in *A Comprehensive Treatise of Electrochemistry*, (Conway B.E.; Bockris J. O´M.; Yeager E.; Khan S. U. M.; White R. E.; Eds.) Plenum Press, London, New York, Vol. 7, pp. 173-239.
[34] Appleby, A. J. (1974). in *Modern Aspects of Electrochemistry*, (Conway B.E., Bockris J. O´M., Eds.) Plenum Press, New York, Vol. 9, pp. 369-389.
[35] Bockris, J. O'M. (2010). in *Electrocatalysis; Computational*, *Experimental and Industrial Applications*, Zinola C.F.; Ed; Vol. 146 CRC Taylor & Francis Group Inc., Chap. 1, pp. 1-42.
[36] Bockris, J. O'M., Reddy, A. K. N., and Gamboa-Aldeco, M. (2000). *Modern Electrochemistry*, 2nd Edition, Plenum Press.
[37] Zinola, C. F. (2010). in *Electrocatalysis; Computational*, *Experimental and Industrial Applications*, Zinola C.F.; Ed; Vol. 146 CRC Taylor & Francis Group Inc., Chap. 9 and 10.
[38] Martins, M. E., Salvarezza, R. C., and Arvia, A. J. (1997). *Electrochim. Acta*, *43*, 549.
[39] Martins, M. E., Zinola, C. F., Andreasen, G., Salvarezza, R. C., and Arvia, A. J. (1998). *J. Electroanal. Chem.*, *445*, 135.
[40] Zinola, C. F. (2022). *Chemelectrochem*, accepted.
[41] Clavilier, J., Faure, R., Guinet, G., and Durand, R. (1980). *J. Electroanal. Chem.*, *107*, 205.
[42] Chen, Q. S., Berna, A., Climent, V., Sun, S. G., and Feliu, J. M. (2010). *Phys.Chem. Chem.Phys.*, *12*, 11407.
[43] Ruthven, D. M. (1984). *Principle of Adsorption and Adsorption Processes*, John Willey and Sons, New Jersey, NJ, USA.
[44] Hill, T. L. (1946). *J. Chem. Phys.*, *14*, 441.

[45] Santos, V. P., and Camara, G. A. (2021). *Res. Surf. Interfaces*, *3*, 100006.
[46] Zinola, C. F. (2011). *Chem.Phys.Chem*, *12*, 172.
[47] Budevski, E., Staikov, G., and Lorenz, W. J. (1996). *Electrochemical Phase Formation and Growth*, p. 17, John Wiley & Sons.
[48] Bockris, J.O'M. (1986). "Dislocations and High Pressure," from P.K. Subramanian *Surface Chemistry and Electrochemistry of Membranes*, A Comprehensive Treatise of Electrochemistry, Vol 7, 383.
[49] Peremans, A., and Tadjeddine, A. (1995). *J. Chem. Phys.*, *103*, 7197.
[50] Santos, E., Quaino, P., and Schmickler, W. (2012). *Phys. Chem. Chem. Phys.*, *14*, 11224.
[51] Shinagawa, T., Garcia-Esparza, A. T., and Takanabe, K. (2015). *Sci. Rep.*, *5*, 1.
[52] Adzic, R. R. (1984). in *Advances in Electrochemistry and Electrochemical Engineering*, J. Wiley & Sons, New York.
[53] Markovic, N. (2002). *Surf. Sci. Rep.*, *45*, 117.
[54] Hamelin, A. (1987). in *Modern Aspects of Electrochemistry*, (Conway B.E.; White R.E.; Bockris J.O.M. Eds.) Vol. 16, Plenum Press, New York, p. 1.
[55] Sluyters-Rehbach, M., and Sluyters, J. H. (1986). *Alternating Current and Pulse Methods*, in *Comprehensive Chemical Kinetics*, (Bamford C.H.; Compton R.G., Eds.) Vol. 26, Chap. 4, pp. 203-354.
[56] Bockris, J. O'M., and Reddy, A. K. N. (*1964*). *Natl. Bur. Std. Misc. Publ.*, *256*, 229.
[57] Diaz, V., Zerbino, J. O., Martins, M. E., Sustersic, M. G., and Zinola, C. F. (2008). *Int. J. Hydrogen Ener.*, *33*, 3502.
[58] Bittins-Cattaneo, B., Cattaneo, E., Königshoven, P., and Vielstich, W. (1991). in *Electroanalytical Chemistry: A Series of Advances*, vol. 17, (Bard A.J., Ed.) Marcel Dekker, New York.
[59] Iwasita, T., and Nart, F. C. (1995). in *Advances in Electrochemical Science and Engineering*, vol. 4, (Gerischer H.; Tobias C.W., Eds.) Verlag Chemie, Weinheim.
[60] Wolter, O., and Heitbaum, J. (1984). *J. Electroanal. Chem.*, *88*, 2.
[61] Wolter, O., and Heitbaum J. (1984). *J. Electroanal. Chem.*, *88*, 6.
[62] Bruckenstein, S., and Gadde, R. R. (1971). *J. Am. Chem. Soc.*, *93*, 793.
[63] Bewick, A., Kunimatsu, K., and Pons, B. S. (1970). *Electrochim. Acta*, *15*, 1345.
[64] Bewick, A., Kunimatsu, K., Pons, B. S., and Russel, J. W. (1984). *J. Electroanal. Chem.*, *160*, 47.
[65] Pons, S. (1983). *J. Electroanal. Chem.*, *150*, 481.
[66] Pons, S., Davison, T., and Bewick, A. (1984). *J. Electroanal. Chem.*, *160*, 63.
[67] Bagotskii, V. S., and Vasiliev, Y. B. (1967). *Electrochim. Acta*, *12*, 1323.
[68] Chandrasekaran, K. C., Wass, J. C., and Bockris, J. O'M. (1990). *J. Electrochem. Soc.*, *137*, 520.
[69] Mac Donald, J. P., Gualtieri, B., Runga, N., Teliz, E., and Zinola, C. F. (2008). *Int. J. Hydrogen Ener.*, *33*, 7048.
[70] Batista, E. A., Malpass, G. R. P., Motheo, A. J., and Iwasita, T. (2004). *J. Electroanal. Chem.*, *571*, 273.
[71] Herrero, E., Franaszczuk, K., and Wieckowski, A. (1994). *J. Phys. Chem.*, *98*, 5074.
[72] Chen, Y. X., Miki, A., Ye, S., Sakai, H., and Osawa, M. (2003). *J. Am. Chem. Soc.*, *125*, 3680.

[73] Rodríguez, J. L., Pastor, E., Zinola, C. F., and Schmidt, V. M. (2006). *J. Appl. Electrochem.*, *36*, 1271.

[74] Parsons, R. (1968). *Discuss. Faraday Soc.*, *45*, 40.

[75] Hogarth, M., Munk, J., Shukla, A., and Hamnett, A. (1994). *J. Appl. Electrochem.*, *24*, 85.

[76] Chrzanowski, W., Kim, H., and Wieckowski, A. (1998). *Catal. Lett.*, *50*, 69.

[77] Sinthika, S., Vala, S. T., Kawazoe, Y., and Thapa, R. (2016). *ACS Applied Materials & Interfaces*, *8*, 5290.

[78] Rodríguez, J. L., Pastor, E., Xia, X. H., and Iwasita, T. (2000). *Langmuir*, *16*, 5479.

[79] Iwasita, T., and Pastor, E. (1994). *Electrochim. Acta*, *39*, 531.

[80] Beden, B., Hahn, F., Lamy, C., Leger, J. M., Tacconi, N., Lezna, R., and Arvia, A. J. (1987). *J. Electroanal. Chem.*, *261*, 215.

[81] Coutanceau, C., Ralotondrainibé, A., Lima, Al., Garnier, E., Pronier, S., Léger, J. M., and Lamy, C. (2004). *J. Appl. Electrochem.*, *34*, 61.

[82] Kolb, D. M. (1978) in *Advances in Electrochemistry and Electrochemical Engineering* (Gerisher H.; Tobias C. W., Eds.) Wiley-Interscience, New York, Vol. 11.

[83] Anderson, A. B., Grantscharova, E., and Seung, S. (1996). *J. Electrochem. Soc.*, *143*, 2075.

[84] Koper, M. T. (2004). *Surf. Sci.*, *548*, 1.

[85] Bonilla, S. H., Zinola, C. F., Rodríguez, J., Díaz, V., Ohanian, M., Martínez, S., and Giannetti, B. F. (2005). *J. Colloid Interface Sci.*, *288*, 377.

[86] Haner, A., and Ross, P. N. (1991). *J. Phys. Chem.*, *95*, 3740.

[87] Bittins-Cattaneo, B., and Iwasita, T. (1987). *J. Electroanal. Chem.*, *238*, 151.

[88] Batista, E. A., Malpass, G., Motheo, A. J., and Iwasita, T. (2004). *J. Electroanal. Chem.*, *571*, 273.

[89] Zinola, C. F. (2019). *J. Solid State Electrochem.*, *23*, 1.

[90] Visintín, A., Canullo, J. C., Triaca, W. E., and Arvia, A. J. (1988). *J. Electroanal. Chem.*, *239*, 67.

[91] Triaca, W. E., Kessler, T., Canullo, J. C., and Arvia, A. J. (1987). *J. Electrochem. Soc.*, *134*, 1165.

[92] Chialvo, A. C., Triaca, W. E., and Arvia, A. J. (1983). *J. Electroanal. Chem.*, *146*, 93.

[93] Zinola, C. F. (2017). *J. Electrochem. Soc.*, *164*, H170.

[94] Arevalo, M. C., Gomis-Bas, C., Hahn, F., Beden, B., Arevalo, A., and Arvia, A. J. (1994). *Electrochim. Acta*, *39*, 793.

[95] Martins, M. E., and Zinola, C. F. (2005). *An. Asoc. Quím. Argent.*, *93*, 4.

[96] Umeda, M., Niitsuma, Y., Horikawa, T., Matsuda, S., and Osawa, M. (2020). *ACS Appl. Energy Mater.*, *3*, 1119.

[97] Zinola, C. F., Gomis-Bas, C., Estiú, G. L., Castro, E. A., and Arvia, A. J. (1988). *Langmuir*, *14*, 3901.

[98] Smolinka, T., Heinen, M., Chen, Y. X., Jusys, Z., Lehnert, W., and Behm, R. J. (2005). *Electrochim. Acta*, *50*, 5189.

[99] Hori, Y., Kikuchi, K., Murah, A., and Suzuki, S. (1986). *Chem. Lett.*, *34*, 897.

[100] Bockris, J. O´M. (1971). *Environment*, *13*, 51.

Chapter 8

The Interplay between RedOx, Photophysics and Surface Process in a Bi_2WO_6 Photocatalyst

Oswaldo Núñez[1,*] and Ronald Vargas[2,3,4,†]
[1] Organic Physicochemistry and Environmental Chemistry Laboratory, Department of Processes and Systems, Chemical Processes Section, Simón Bolívar University, Caracas, Venezuela
[2] Chascomús Technological Institute, National Council for Scientific and Technical Research, Chascomús, Argentina
[3] School of Bio and Nanotechnologies, National University of San Martín, Chascomús, Argentina
[4] Department of Chemistry, Simón Bolívar University, Caracas, Venezuela

Abstract

Light-induced electron transfer from Bi_2WO_6 to an electron acceptor in the electrolyte solution, is explained using the interaction between the conduction band potential (E_{CB}) and the density of states (DOS) of the *RedOx* couple in the electrolyte. Energy diagrams were used to illustrate the thermodynamic and kinetic concepts involved in E_{CB}-DOS interactions. In Bi_2WO_6, E_{CB} is +0.5 V *vs.* NHE, and the reduction potential of $O_2/O_2^{\bullet-}$ is -0.33 V *vs.* NHE, resulting in a negative potential difference for electron transfer. Therefore, further $O_2^{\bullet-}/H_2O_2$ reduction at +0.89 V *vs.* NHE is the thermodynamic driving force that allows the quenching process. Thus, photooxidation rates using Bi_2WO_6 are limited

* Corresponding Author's Email: onunez@usb.ve.
† Corresponding Author's Email: ronaldvargas@intech.gov.ar.

In: A Closer Look at Chemical Kinetics
Editor: Victor Martinez-Luaces
ISBN: 979-8-88697-484-3

by the formation of adsorbed hydrogen on the surface, that implies an electron transfer from the E_{CB} to the surface *RedOx* couple Bi_2WO_6,H^+/ Bi_2WO_6-H (-0.6 V *vs.* NHE).

This large negative potential is overcome by light excitation and/or electrochemical assistance and the surface phenomenon is consistent with previous kinetic isotope effect results. The relevance of the quenching process between dissolved O_2 and the excited semiconductor is shown using the reduction potentials of O_2 and the E_{CB} values. A fraction of the electrons excited is transferred to O_2 instead of emitting light upon returning to the valence band, as is the case when O_2 is present at a low concentration. The steady-state photoluminescence results of the Bi_2WO_6-Water suspensions indicate that emission light intensity is decreased when O_2 is present. Now, the modification of Bi_2WO_6 with 4-stilbenecarboxaldehyde (4SCA) opens the possibility of delocalizing photogenerated electrons through the molecule anchored on the semiconductor. This delocalization occurs at potentials lower than the E_{CB} of Bi_2WO_6, so the changes in the deactivation kinetics are expected as well as tuning from the semiconductor modification protocol. The time-resolved photoluminescence results measured using Bi_2WO_6 and Bi_2WO_6-4SCA support the described kinetic effect.

The arguments discussed in this chapter help in understanding the interplay of the physical and chemical phenomena which condition *RedOx* reactions on illuminated semiconductor surfaces.

Keywords: photocatalysis, kinetics, electron transfer, photoluminescence, quenching.

Introduction

Heterogeneous photocatalysis is an interesting strategy to carry out chemical reactions induced by solar energy, this through a semiconductor material. Thus, it is important to rationalize the kinetic and thermodynamic behavior of photocatalytic systems [1]. In fact, we have been involved with semiconductors research working on oxidation and mineralization (CO_2 + H_2O) of organic contaminants by carrying out kinetic studies to understand the phenomena, elucidate chemical behaviors and visualize project applications [2-13].

The interpretation of kinetic data using Langmuir-Hinshelwood (LH) type models [1, 14-16] has been key to the further development of research in photocatalysis. The basic form of the LH equation is $r = kKc/(1 + Kc)$, where r is the reaction rate, c is the concentration of organic compound, K is the organic compound, semiconductor adsorption, desorption equilibrium constant and k is the reaction rate at maximum organic compound coverage. Now, it is common to check the linearity of $1/r$ *vs.* $1/c$ data to analyze whether photocatalytic systems can be described kinetically with the LH equation [1, 15-16]. Lessons and strategies have been reported from previous works to improve and understand the organic compounds degradation processes [1, 5-6, 15], which have revealed that improvements in chemical kinetics are fundamental, and have also allowed the decoupling of kinetic effects, summarized in the values of k, and the thermodynamic effects, gathered in the values of K. However, to delve into the causes that determine the photocatalytic processes and the experimental determination of these kinetic parameters, we insist on detailing the thermodynamic considerations that determine the kinetics of the RedOx processes during photocatalysis.

The understanding of the RedOx process at the semiconductor-electrolyte interface became crucial for the objective of optimizing the photocatalytic performance. However, we found a vocabulary and knowledge gap between the fields of solid-state chemistry, electrochemistry, and physical organic chemistry (POC) in solution. In this manuscript, we try to fill this gap in a comprehensive way. Our approach is intended to explain the semiconductor-electrolyte results from a POC approach where energy diagrams, driving force, product development control, isotope effect, and others POC concepts are applied to improve and explain the electron transfer process. We hope this manuscript finally contributes to a better understanding of these processes and consequently improves the optimization of the variety of semiconductor applications.

It is advantageous to use a photocatalyst such as Bi_2WO_6 to illustrate the POC approach described above. Bi_2WO_6 is the simplest Aurivillius oxide [17], exhibits *n*-type semiconductor behavior [18], and excellent photocatalytic performance in processes oriented to both wastewater treatment [17-18] and energy applications [12, 17]. Now, there is a fundamental aspect that determines its experimental behavior and that was recently pointed out in the scientific literature by our group; this is the relatively positive potential value of the CB of Bi_2WO_6 (+0.5 V *vs.* NHE) [11]. Therefore, there is a natural limit to the magnitude of the driving force that conditions the kinetics of the process involved during the photocatalytic process. Many reports have focused on the modification of Bi_2WO_6 with other materials to improve its photocatalytic properties [19-21]. However, knowing how the driving force is limited, more appropriate strategies can be designed based on the enhancement of electron transfer that occurs in this semiconductor.

In this work, some key physicochemical aspects about electron transfer at semiconductor-electrolyte interfaces will first be discussed, emphasizing the use of energy diagrams to understand possible thermodynamic phenomena and link them with their kinetics. In order to build a map to understand the interrelation of *RedOx*, photophysical, and surface phenomena that take place at an illuminated Bi_2WO_6|Water interface, selected spectroscopic, electrochemical, and kinetic results related to the behavior of the Bi_2WO_6 photocatalyst have been juxtaposed in a holistic interpretation.

Potentials and Density of States (DOS) Interactions: Energy Diagrams and Electron Transfer

In *n*-type semiconductors such as TiO_2 and Bi_2WO_6, the majority of the carriers are electrons; therefore, the Fermi level, E_F, or work required to add one electron to the semiconductor (see Figure 1A), is at lower semiconductor absolute energy E and closer to the conduction band. Lower absolute energy E (eV) implies more negative standard potential $E°$ since $E = -4.5 \text{ eV} - E°$ [22]. The conduction band (CB) potential, E_{CB}, and the valence band potential (VB), E_{VB}, correspond to the lowest energy level of the CB and the highest energy level of the VB, respectively. The concept of VB and CB levels is grossly interpreted as the lower unoccupied molecular orbital (LUMO) and highest occupied molecular orbital (HOMO) in molecules, but VB and CB are energy bands that overlap many molecular orbitals instead of single states. Above CB there is a continuous energy band that can be reached by electrons using light excitement or external bias polarization; similarly, below VB there is a continuous energy band with electrons.

The difference between E_{VB} and E_{CB} is the band gap (E_g): $E_g = E_{VB} - E_{CB}$ [23]. In principle, there is no energy level between E_{VB} and E_{CB}, this is precisely what turns these solids into semiconductors. Now, E_F changes in energy according to the equilibrium potential of the *RedOx* couple at the electrolyte in contact with the semiconductor, so, VB and CB suffer flattening and bending. For instance, if the E_F increases in energy, which is more negative standard potential and lower semiconductor absolute energy, the bands bend downward and electron enrichment at the semiconductor interface occurs (Figure 1B). On the contrary, if the equilibrium potential of the *RedOx* couple at the electrolyte is more positive than E_F, bands bend upward, and electron depletion occurs at the semiconductor interface (Figure 1C) [22-23].

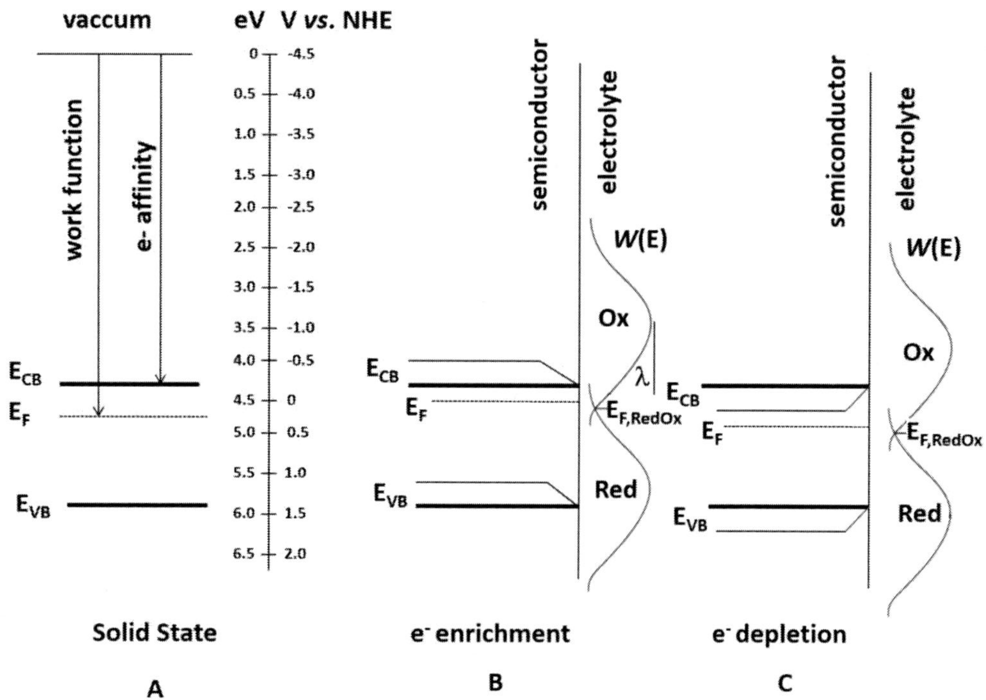

Figure 1. Energy diagrams. A. Terms definition; bands positions before contact with the solution, absolute energy in eV (solid state) and standard potential $E°$ (solution) in V *vs*. NHE; B. Band bending when E_F is more negative than equilibrium potential (A): electron enrichment at the interface occurs; C. Band bending when E_F is more positive than equilibrium potential (A): electron depletion occurs at the interface. In the electrolyte: Ox + e^- = Red, $E_{F,RedOx}$ is the Fermi level of the *RedOx* couple, $W(E)$ is a function representing the electronic levels or wave distribution function, and λ is the solvent reorganization energy. *NHE*: normal hydrogen electrode.

Changes in the electron density at the interface influence the rate of electron transfer to the semiconductor surface [23]. Since the conductivity at the semiconductor is lower than the solution one, electron transfer rates mainly depend on the electron density at the surface, which takes place, especially under spontaneous exergonic conditions as there is depletion of electrons at the interface as shown in Figure 1C. If the following RedOx reaction is considered to occur Ox + e- = Red (Ox is an electron acceptor and Red and electron donor), then some kinetic considerations must be invoked:

1. The electron transfer rate constant measured under this condition is *pseudo*-first order, since electron concentration ([e]) is lower than concentration of the RedOx couple ([Red] and/or [Ox]), that is: [e] < [Ox].
2. In the electrolyte, a *RedOx* pair wave distribution function, $W(E)$, defines the density of state (DOS), that is: DOS(E) = [Ox] x $W(E)$.

The resulting $W(E)$ function also affects the electron transfer rate. Although the electron transfer is fast, some delay occurs because of the solvent reorganization energy (λ) after the electron transfer at the interface. This solvent reorganization energy was assertively identified by Marcus [24-25], and is given by the difference between the energy level of Ox (E_{Ox}) and Fermi level of the *RedOx* couple ($E_{F,RedOx}$) illustrated in Figure 1B. The larger λ the wider is DOS. The distance increment between electron donor and electron acceptor makes λ also larger. It should be noted that $W(E)$ component of DOS shown in figures is defined by the Gaussian function [23]: $\exp[-(E - E_{F,RedOx} - \lambda)^2/(4kT\lambda)]$.

If $W(E_F)$ is more negative than the E_{CB}, then electron transfer does not occur spontaneously due to the unfavorable thermodynamic energy change (Figure 2A). There is no DOS of Ox available to interact with E_{CB}. However, if it is thermodynamically favorable: $W(E_F) > E_{CB}$, the electron transfer from E_{CB} to Ox in the electrolyte occurs (Figure 2B). Even under this condition, the electron transfer rate from the semiconductor will depend on the $W(E)$ alignment with E_{CB}, where a maximum rate occurs (k_2) at the oxidant reduction potential value (E_{F2} in Figure 2B). Alignment to less positive potential (E_{F1} in Figure 2B) decreases the electron transfer rate (k_1) as predicted by Marcus normal zone [24-25], but also at even more positive potential (E_{F3} in Figure 2B) a decrease in the electron transfer rate may occur due to the Marcus inverted zone (k_3) [24-25].

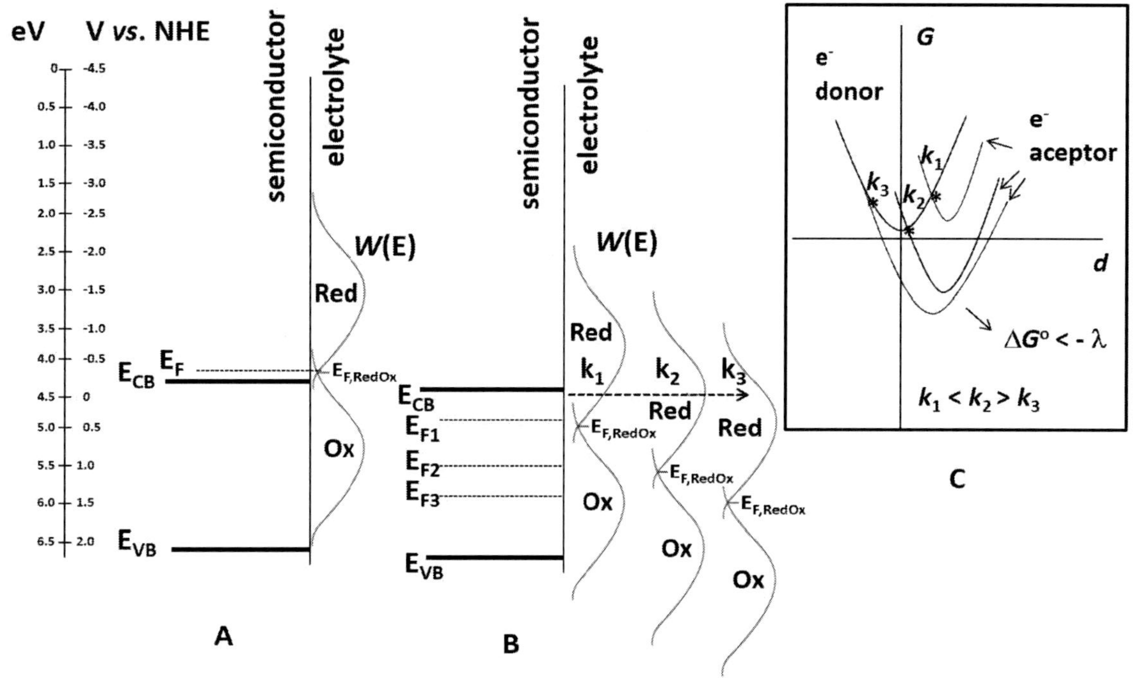

Figure 2. Energy diagrams. A. E_F less positive than E_{CB}: electron transfer does not occur due to the lack of effective interaction between E_{CB} and Ox DOS; B. E_F more positive than E_{CB}: electron transfer occurs at different rates ($k_1 < k_2 > k_3$) depending on the Ox DOS alignment with E_{CB}; C. G *vs.* d plot and transition states (*) associated with the rates k_1, k_2 and k_3, according to Eyring equation adapted to the Marcus approach.

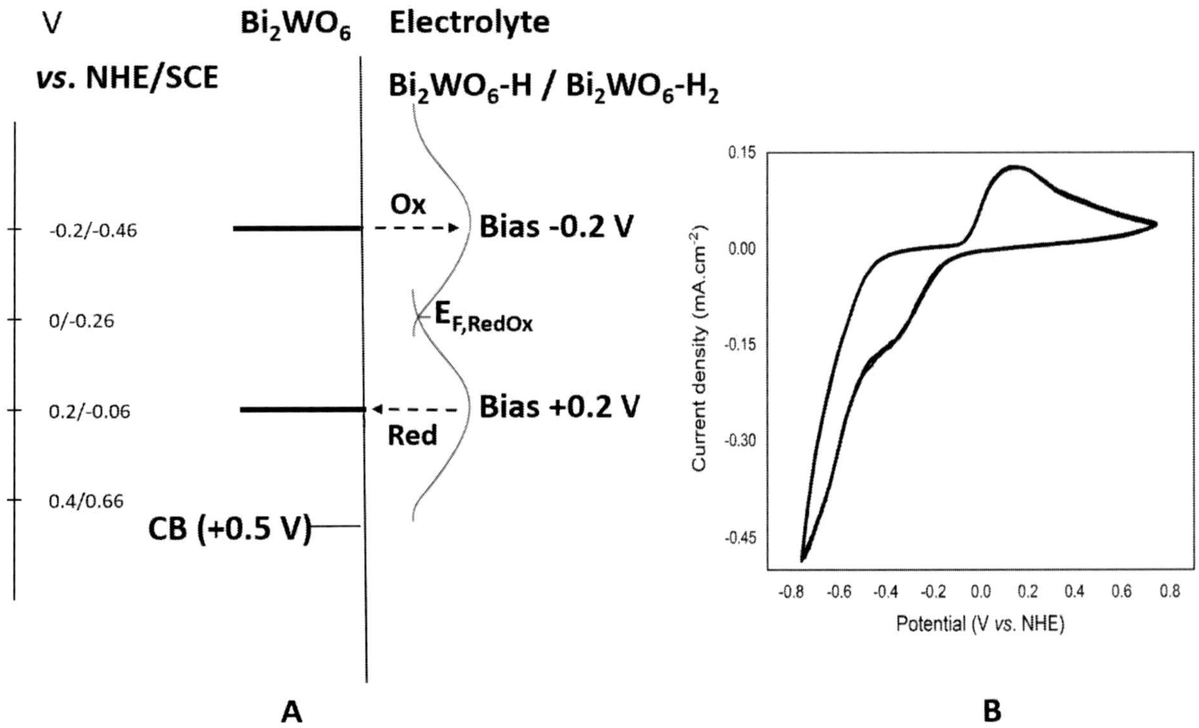

Figure 3. A. Energy diagram depicting the Ox and Red maximum interaction with the bias value imposed; B. Cyclic voltammetry of Bi_2WO_6 electrode at pH = 7 and 10 mV/s in 0.1 M Na_2SO_4 (cycle 20) (see the text). *SCE*: standard calomel electrode.

While reaction rates typically increase with increasing exergonicity, electron transfer reactions should, according to Marcus theory, become slower with a very negative $\Delta G°$ value when $\Delta G° < -\lambda$. As shown in Figure 2C and using the Gibbs free energy (G) *vs.* distance (d) plot, a higher transition state (*) for k_3 than the corresponding one (*) to k_2 might be reached in the case of very exergonic reactions. Therefore, $k_1 < k_2 > k_3$ as expected for the reaction rate using the Eyring equation adapted with the Marcus approach: $k = (k_B T/h).\exp[-(\lambda + \Delta G°)^2/(4\ \lambda\ RT)]$ [26]. Notice that the numerator into the exponential term, which is: $(\lambda + \Delta G°)^2$, becomes higher in the case of k_3 ($\Delta G° < -\lambda$), than in the case of k_2, then the rate constant decreases according to the discussed equation.

Regarding the energy diagrams, it is important to indicate that to maintain a single axis as a reference (vertical one), showing semiconductor and electrolyte E values, DOS are drawn parallel to the vertical axis as well as the semiconductor-electrolyte interface. If DOS are drawn perpendicular (interface rotation to the left), the figure will be equivalent, but DOS and potentials values now must be shown on a horizontal-axis; that is, semiconductor E values, CB for instance, shown on the vertical-axis and the electrolytes ones on a horizontal-axis. This change would complicate figures where E_F, E_{CB} and E_{VB} directly interact with DOS when electron transfer occurs and only one vertical axis having all E values is more convenient. However, visualizing this vertical change helps to understand Franck-Condon principle in which electron excitement from VB ends up at the corresponding DOS vibrational level of the RedOx couple in solution without nuclei movement. This movement, required for solvation of the reduced Ox species at the solution after the electron transfer, causes a wider DOS decreasing its intensity and reducing the electron transfer rate. In Figure 2 instead of the last direct vertical transfer, E_{CB}-DOS alignments are shown with horizontal arrows.

If the semiconductor is attached to an electrode and external potential is applied (bias polarization), E_F still depends on the electrolyte DOS, that is, it is related to the oxidized and reduced species. However, since the applied energy is greater than E_{CB}, DOS alignment is given by the bias as shown in the energy diagram shown in Figure 3A. To illustrate the point, Figure 3B presents a cyclic voltammetry of a Bi_2WO_6 electrode obtained according to the protocol reported previously [11]. The electrochemical response is due to the following *RedOx* couple on the surface: Bi_2WO_6-H/Bi_2WO_6-H_2. The two signals at -0.2 V *vs.* NHE and +0.2 V *vs.* NHE correspond to the interaction of the bias with the Ox (Bi_2WO_6-H_2) reduction potential and the Red (Bi_2WO_6-

H) reduction one. In addition, it is commented that recently experimental and computational studies on nanostructured electrocatalysts were reported, through which an unprecedented correlation was established between electrocatalytic activity and the orbital composition that defines the DOS [27].

In all cases, and for future discussions, it is important to recognize that if a bias is applied, then a new equilibrium is established between the semiconductor and the electrolyte. That is, a new potential value is defined at the interface: E = bias, and, in principle, this agrees with the Nernst equation: E = bias = $E° - (RT/nF) \ln ([Red]/[Ox])$. If bias is not present, E_F is given by the RedOx equilibrium value. However, since the interaction of DOS is with E_{CB} (or with more negative potentials), only if the proper energy is applied, e.g., polarization and/or radiation with $h\nu > E_g$, electron transfer can occur.

Light Excitation of Semiconductors and RedOx Reactions

When light irradiates the semiconductor surface at $h\nu = E_g$, an electron is excited from E_{VB} to E_{CB} and a charge carrier pair is formed, that is, an electron (e^-) at E_{CB} and a hole (h^+) at E_{VB}. Now, the energy diagram of Bi_2WO_6 will be used to illustrate the selected phenomena, see Figure 4.

If $h\nu > E_g$, an electron from VB is also excited to a higher energy level in the CB continuum and can decay to E_{CB} by giving up energy to the surroundings unless there is a thermodynamically and kinetically favored reduction reaction. In all cases, as long as the excited electron is transferred from CB to any acceptor species in the electrolyte as the dissolved O_2, a hole in VB oxidizes a compound at the interface. In fact, the oxidation of water to form hydroxyl radicals ($HO^•$) is proven in many semiconductors [28-31]. Thus, organic compounds can be oxidized directly in the VB and/or indirectly by the attack of $HO^•$. Therefore, degradation/mineralization (CO_2 + water) of the organic matters (OM) can occur after consecutive oxidation steps of the organic compound and the intermediates (first oxidation product represented as OM-OH) [30-31].

If the light is ON and the n-type semiconductor is connected to a collector electrode forming a photoanode and externally connected to another electrode serving as a cathode, then a reduction reaction can occur at the suitable cathode and oxidation reactions at the formed semiconductor electrode [30]. If the E_{CB} of the semiconductor electrode is more negative than the standard *RedOx* potential of the oxygen reduction reaction, then a current flow throws off the external circuit and electrical energy is produced in the formed photocatalytic fuel cell [31].

If $E_{CB} > E°$ of any electron acceptor, then the photocatalytic process is not spontaneous. However, the reduction reaction can be externally assisted. In fact, applications on the use of electrochemical assistance of photocatalysis to promote water reduction reaction and hydrogen (H_2) photosynthesis are reported in the literature [12, 30-33].

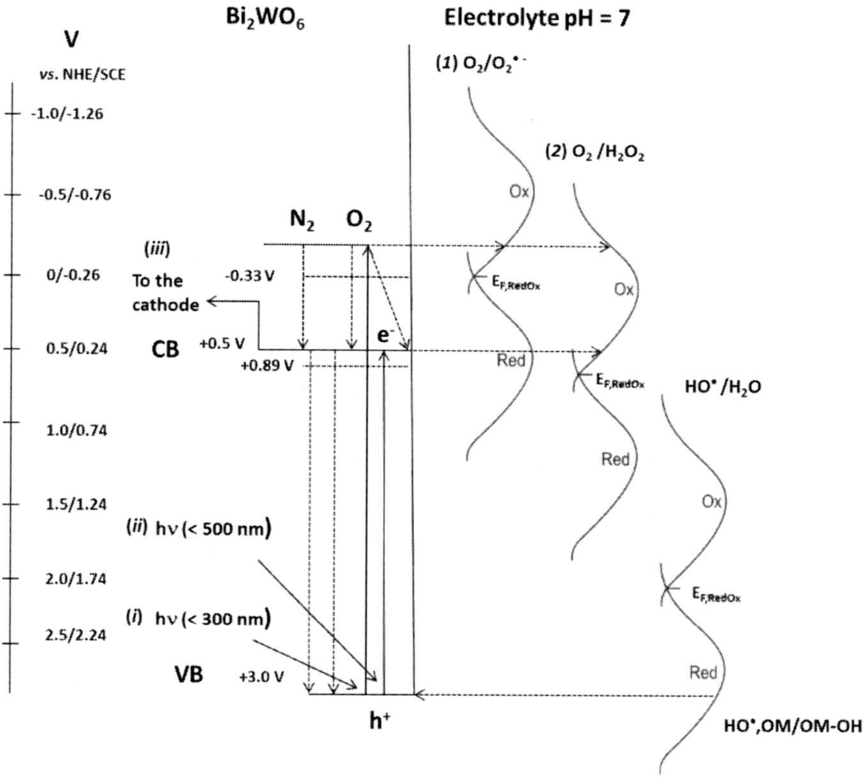

Figure 4. Energy diagram of Bi_2WO_6 in water at pH = 7. Electron (e^-) is excited to an energy level > E_{CB} leaving a hole (h^+) at the VB. Light excitation at (*i*) $hv < 300$ and (*ii*) < 500 nm respectively are indicated. Excited electron can be transferred to an acceptor species in solution via its $E_{F,RedOx}$-DOS interaction, for example: (*1*) reduction of O_2 to $O_2^{\cdot-}$ and (*2*) reduction of O_2 to H_2O_2. Simultaneously, water is oxidized at the VB ($HO^{\cdot}/H_2O$) and at more positive potential OM is oxidized ($HO^{\cdot}$,OM/OM-OH). If O_2 is not present (by supplying an atmosphere of N_2 after bubbling the solution), the electron returns to CB from where it goes down to the VB emitting light at $hv = E_g$, or it could be transferred to the cathode (*iii*) causing current to flow through the external circuit.

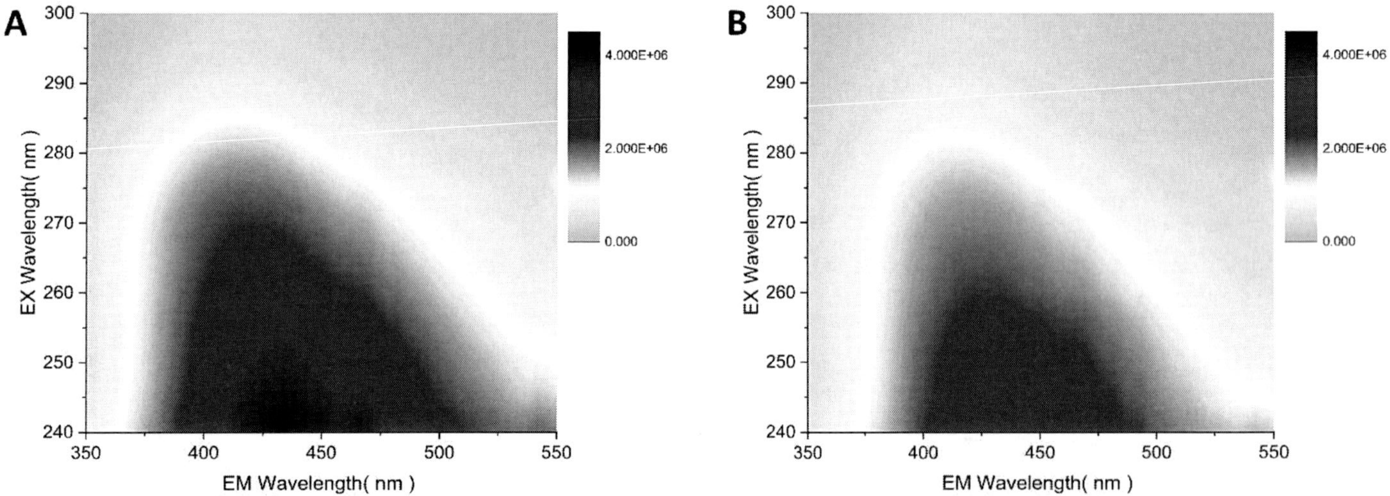

Figure 5. EEM of Bi_2WO_6 water suspension at pH = 7 according to the reported protocol [12-13]. A. In presence of N_2 and B. In presence of O_2. Light intensity (*I*): $I_A > I_B$, for example: at an excitation wavelength of 240 nm, $I_B/I_A \sim 0.78$ is obtained at an emission wavelength of 440 nm.

Recombination by Light Emission

The process that limits photocatalysis is charge carrier recombination, which occurs when the lifetime of the electron and hole is short, and these carriers can combine to emit light or dissipate energy as heat [34-35]. For example, Figure 5 shows photoluminescence spectroscopy in Bi_2WO_6 colloidal suspension results, specifically excitation – emission matrix (EEM). Light excitation (EX on the vertical axis) occurs at the bandgap energy but also at continuous energy levels above the bandgap; however, light emissions (EM on the horizontal axis) occur, at principle, from the bandgap, but emissions at energies below E_g are also revealed, these due to surface states [13].

At this point we will remark that in the analysis of the energy diagrams we have not discussed the superficial or mid-gap states; these energy states are due to defects in the semiconductors, doping and discontinuities in the crystal lattice [23]. Briefly, the surface state introduces distributed and/or localized states in DOS. Therefore, these states can alter the course of photophysical and *RedOx* processes that take place at the interface. Two types of surface states will be considered. The first are the so-called surface traps, which have lower but closer energy than CB and behave as distributed energy states. The second are called deep traps that are energetically located between the VB and the CB, in addition to behaving as a monoenergetic state [23, 36-37].

According to Figure 5A, if O_2 is absent, there is no *RedOx* quenching the excited electron, and emission via electron/hole recombination occurs and the light intensities of the signals are higher. However, Figure 5B indicates that when O_2 is present, light emissions also occur but at quite lower intensities due to the electron quenching via O_2 reduction. This experiment proves that a *RedOx* reaction can occur even in the case that $E_F < E_{CB}$ (Figure 2A) if light ($h\nu > E_g$) irradiates the semiconductor surface (Figure 4).

Now, if light irradiates the semiconductor surface, a steady-state concentration of electrons and holes is established meanwhile light is ON. In this case, the light energy emitted is weak and its wavelength is determined by the photophysical process that takes place due to the DOS semiconductors [34]. Quantitative models have been reported indicating that photoluminescence at semiconductor-electrolyte interfaces is inverse to photocurrent (electron transfer) [38], precisely which helps support the ideas and evidence here discussed.

The steady-state photoluminescence studies of Bi_2WO_6-Water suspensions (Figure 5) indicate that the intensity of the emission light decreases when O_2 is present [12]. In order to analyze the kinetics of photophysical processes involving charge carriers, we will first consider a discussion of previously published results by our group on the kinetics of photoluminescence intensity decay in Bi_2WO_6 suspensions [12]. Then, we will discuss the interface modification strategy to affect decreased useful life [13].

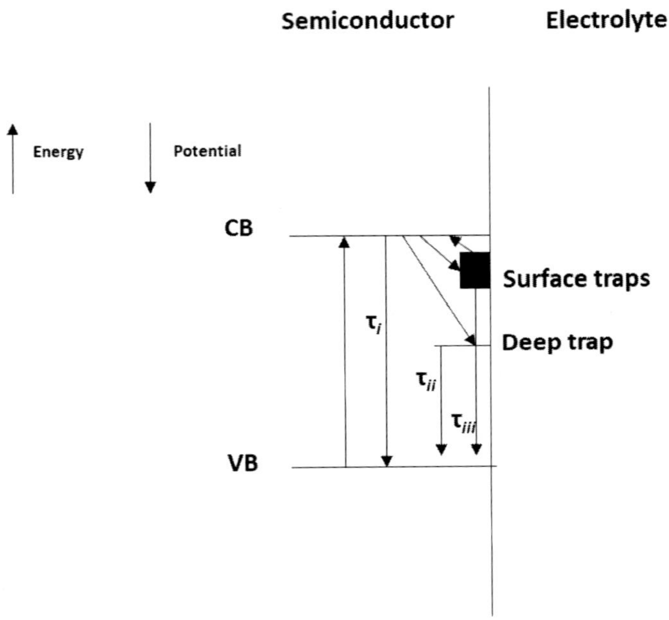

Figure 6. Energy diagram indicating the superficial states and the recombination lifetimes (towards the VB): τ_i (CB), τ_{ii} (superficial trap) and τ_{iii} (deep trap).

According to the DOS theory of the semiconductor-electrolyte interface [34, 36], the radiation emission can be explained considering three emission states [12-13], these would be: (*i*) electron decay from CB, (*ii*) electron decay from surface traps with lower but closer energy than CB (shallow traps) and (*iii*) electron decay from deep trap located between CB and VB. In the energy diagram shown in Figure 6, both shallow and deep traps are indicated at the semiconductor-electrolyte interface; in addition, the lifetimes of the recombination processes described above are associated. For Bi_2WO_6|Water suspension in an N_2-saturated atmosphere, the following lifetime values have been reported: τ_i = 6.5 ns, τ_{ii} = 14.7 ns, and τ_{iii} = 3.0 ms, respectively. On the other hand, in the presence of O_2, the lifetimes were: τ_i = 3.1 ns, τ_{ii} = 10.6 ns and τ_{iii} = 168 ns, respectively [12]. When comparing the magnitudes of the lifetimes in an atmosphere saturated with N_2 and in the presence of O_2, it is observed that the photophysical deactivation processes are affected by the presence of O_2, being consistent with the deactivation of electrons in the CB by the quenching process with the dissolved O_2 that improves electron decay by this additional mechanism.

Now, as reported [13], the modification of Bi_2WO_6 with molecules such as 4-stilbenecarboxaldehyde (4SCA) has opened the possibility of delocalizing photogenerated electrons through the molecule anchored in the semiconductor, simulating a "molecular wire". This delocalization occurs at potentials more positive than the E_{CB} of Bi_2WO_6, so differences in the deactivation kinetics are expected and adjusted from the semiconductor modification protocol. In fact, reported time-resolved photoluminescence results for Bi_2WO_6 with 20% 4SCA on the surface in N_2 atmosphere yielded the following lifetime values: τ_i = 4.06 ns, τ_{ii} = 11.50 ns, and τ_{iii} = 0.12 μs, respectively [13]. All lifetimes for the Bi_2WO_6|Water system are shorter than those reported for the Bi_2WO_6-4SCA|Water system due to the stabilization induced by the electron delocalization. These kinetic results allowed us to show that the variations in the deactivation kinetics can be adjusted from the semiconductor modification protocol, being able to control and eventually design strategies for the transfer of electrons from the CB to an acceptor molecule in solution. It is interesting to postulate that this type of strategy is compatible with improved Bi_2WO_6 surfaces [39-40], being able to explore the physicochemical phenomena that occur in photocatalytic systems to direct efforts towards challenging photo-*RedOx* reactions [12, 17, 41-43].

Conclusion

In short, *n*-type semiconductors such as TiO_2 and Bi_2WO_6 interact with the *RedOx* couple in the electrolyte transferring its majority charge carrier (e^-) from the conduction band (E_{CB}) and the minority charge carrier (h^+) from the valence band (E_{VB}). These transference reactions are simultaneous but one of them limits the overall process, that is, $d[h^+]/dt = d[e^-]/dt$. Comparing the potentials of the bands with the Fermi level of the *RedOx* couple ($E_{F,RedOx}$): (*i*) $E_{F,RedOx} - E_{CB}$ and (*ii*) $E_{VB} - E_{F,RedOx}$, determines which band (CB or VB) limits and defines the driving force for photocatalysis. When E_{CB} is more positive, as is the case of Bi_2WO_6, $E_{F,RedOx} - E_{CB}$ is lower (less driving force), and the electron transfer from the semiconductor to the electron acceptor limits the overall process.

Now, since the alignment between E_{CB} and DOS depends on the equilibrium potential of the *RedOx* couple, the more positive the $E_{F,RedOx}$ value, the better the interaction with the E_{CB} up to a point where a maximum is reached. Further positive $E_{F,RedOx}$ value may reduce the electron transfer rate; this is according to the Marcus inverted kinetic zone. When a bias is applied, the electron acceptor interacts with the actual bias value across the interface. Since the bias potential is more negative than E_{CB}, the electron enrichment at the interface makes electron transfer faster and less selective. In all cases, electron transfer is involved in a delicate balance with recombination, so it is important to note that to favor electron transfer, recombination pathways and their kinetics must be discouraged. This can be achieved from the synthesis of materials with improved optoelectronic properties and the use of electrochemical assistance to extract photogenerated electrons.

Acknowledgments

We thank the Environmental Management Unit of the USB (Venezuela), TWAS and TYAN for their financial support, and especially Dr. Franco M. Cabrerizo from INTECH-CONICET/UNSAM (Argentina) for his hospitality during the visit of one of us (RV) where part of this manuscript was written.

References

[1] Liu, B., Zhao, X., Terashima, C., Fujishima, A., and Nakata, K. (2014). *Phys. Chem. Chem. Phys.*, 16, 8751-8760.
[2] Barrios, N., Sivov, P., D'Andrea, D., and Núñez, O. (2005). *Int. J. Chem. Kinetics.*, 37, 414-419.
[3] Kuehr, I., and Núñez, O. (2007). *Pest. Manag. Sc.*, 63, 491-494.
[4] Vargas, R., and Núñez, O. (2008). *J. Mol. Catal. A: Chem.*, 294, 74-81.
[5] Pardo, G., Vargas R., and Núñez, O. (2008). *J. Phys. Org. Chem.*, 21, 1072-1078.
[6] Vargas, R., and Núñez, O. (2009). *J. Mol. Catal. A: Chem.*, 300, 65-71.
[7] Vargas, R., and Núñez, O. (2010). *Solar Energy.*, 84, 345-351.
[8] Vargas, E., Vargas, R., and Núñez, O. (2014). *Appl. Catal. B: Environ.*, 156-157, 8-14.
[9] Nuñez, O., Rivas, C., and Vargas, R. (2014). *J. Phys. Org. Chem.*, 28, 191-198.
[10] Nuñez, O., Rivas, C., and Vargas, R. (2017). *J. Phys. Org. Chem.*, 30, e3659.

[11] Núñez, O., Madriz, L., Carvajal, D., Tatá, J., and Vargas, R. (2019). *J. Phys. Org. Chem.*, 32, e3952.

[12] Madriz, L., Tatá, J., Carvajal, D., Núñez, O., Scharifker, B. R., Mostany, J., Borras, C., Cabrerizo, F. M., and Vargas, R. (2020). *Renewable Energy.*, 152, 974-983.

[13] Madriz, L., Parra, M., García Einschlag, F. S., Núñez, O., Cabrerizo, F. M., and Vargas, R. (2021). *J. Phys. Chem. C.*, 125, 12682-12689.

[14] Braslavsky, S. E., Braun, A. M., Cassano, A. E., Emeline, A. V., Litter, M. I., Palmisano, L., Parmon, V. N., and Serpone, N. (2011). *Pure Appli. Chem.*, 83, 931-1014.

[15] Ollis, D. F. (2018). *Front. Chem.*, 6, 378.

[16] Loddo, V., Camera Roda, G., and Parrino, F. (2019). In *Heterogeneous Photocatalysis: Relationships with Heterogeneous Catalysis and Perspectives;* Marcí, G., Palmisano, L., Eds., Elsevier: Amsterdam, Netherlands, Vol. 1, pp. 215-233.

[17] Li, W., and Zhu, Y. (2020). In *Surface Science of Photocatalysis;* Yu, J., Jaroniec, M., Jiang, Ch., Eds., Interface Science and Technology Book Series, Elsevier: Amsterdam, Netherlands, Vol. 31, pp. 379-413.

[18] Alfaifi, B. Y., Tahir, A. A., and Upul Wijayantha, K. G. (2019). *Solar Eng. Mater. Solar Cells.*, 195, 134-141.

[19] Liu, X., Gu, S., Zhao. Y., Zhou, G., and Li, W. (2020). *J. Mater. Sci. Technol.*, 56 45-68.

[20] Orimolade, B. O., Idris, A. O., Felini, U., and Mamba, B. (2021). *Environ. Poll.*, 15, 117891.

[21] Zhu, Z., Wan, S., Zhao, Y., Qin, Y., Ge, X., Zhong, Q., and Bu, Y. (2021). *Nano Select.*, 2, 187-215.

[22] Memming, R. (2015). *Semiconductor Electrochemistry*; Wiley-VCH: Weinheim, Germany, Vol. 2, pp 1-470.

[23] Sato, N. (1999). *Electrochemistry at Metal and Semiconductor Electrodes*; Elsevier: Amsterdam, Netherlands, Vol. 1, pp. 1-413.

[24] Marcus, R. A., and Sutin, N. (1985). *Biochim. Biophys. Acta.*, 811, 265-322.

[25] Stubbe, J., Nocera, D. G., Yee, C. S., and Chang, M. C. Y. (2003). *Chem. Rev.*, 103, 2167-2201.

[26] Alkhalifah, M. A., Howchen, B., Staddon, J., Celorio, V., Tiwari, D., and Fermín, D. J. (2022). *J. Am. Chem. Soc.*, 144, 4439-4447.

[27] Schneider, J., Matsuoka, M., Takeuchi, M., Zhang, J., Horiuchi, Y., Anpo, M., and Bahnemann, D. W. (2014). *Chem. Rev.*, 114, 9919-0086.

[28] Singh, P., Abdullah, M. M., Sagadevan, S., and Ikram, S. (2019). In *Photocatalysis: Perspective, Mechanims, and Applications*; Singh. P., Abdullah. M. M., Ahmad. M., Ikram. S., Eds., Nanotechnology Science and Technology Book Series, Nova Science Publishers, New York, USA, Vol. 1, pp. 1-18.

[29] Lianos, P. (2017). *App. Catal. B: Environ.*, 210, 235-254.

[30] Espinoza-Montero, P., Vargas, R., Alulema-Pullupaxi, P., and Fernandez, L. (2022). In *Advanced Oxidation Processes for Wastewater Treatment: An Innovative Approach;* Shah, M. P., Dera, S. P., Tildiz. G., Eds., Recent Trends in Water Research Book Series, CRC: Florida, USA, Vol. 1, pp. 53-68.

[31] Adamopoulos, P. M., Papagiannis, I., Raptis, D., and Lianos, P. (2019). *App. Catal. B: Environ.*, 9, 976.

[32] Sun, D., Fan, W., Wang, F., Bai, Y., Bai, H., and Shi, W. (2022). *Int. J. Hydroge. Eng.*, 47, 940-949.

[33] Gilliland, D. D. (1997). *Mater. Sci. Eng.*, R18, 99-400.

[34] Qian, R., Zong, H., Schneider, J., Zhou, G., Zhao, T., Li, Y., Yang, J., Bahnemann, D. W., and Pan, J. H. (2019). *Catal. Today.*, 335, 78-90.

[35] Bisquert, J., Zaban, A., Greenshtein, M., and Mora-Seró, I. (2004). *J. Am. Chem. Soc.*, 126, 13550-13559.

[36] Monllor-Satoca, D., Díez-García, M. I., Lana-Villareal, T., and Gómez, R. (2020). *Chem. Comm.*, 56, 12272-12289.

[37] Smandek, B., Chmiel, G., and Gerischer, H. (1989). *Ber. Bun. Phys, Chem.*, 93, 1094-1103.

[38] Zhou, Y., Zhang, Y., Lin, M., Long, J., Zhang, Z., Lin, H., Wu, J. C. S., Wang, X., Chung, H. Y., Wu, X., Amal, R., and Ng, Y. H. (2020). *Mol. Catal.*, 487, 110887.

[39] Madriz, L., Carrero, H., Herrera, J., Cabrera, A., Canudas, N., and Fernández, L. (2011). *Top. Catal.*, 54, 236-243.

[40] Nunez, O., Sattayamuk, D., Saelee, T., Yamashita, H., Kuwahara, Y., Mori, K., Praserthdam, P., and Praserthdam, S. (2021). *Chem. Eng. J.*, 409, 128141.

Chapter 9

Structure–Activity Relationship of Functionalized Porous Graphene Oxide Nanocomposites for Catalytic Applications

Amit Dubey[1,*], Savita Verma[3] and C. Sasikumar[2]

[1] Department of Chemistry,
Maulana Azad National Institute of Technology (MANIT), Bhopal, (M.P.) India
[2] Department of Materials and Metallurgical Engineering,
Maulana Azad National Institute of Technology (MANIT), Bhopal, (M.P.) India
[3] Department of Chemistry,
Government Science College, Jabalpur, (M.P.) India

Abstract

Functionalized graphene oxide (FGO) with formamidine sulfinic acid (FSA) was synthesized by post functionalization method for studying antibacterial activity against gram-positive and gram-negative bacteria, particularly for *S. mutans* and *Pseudomonas* bacteria. The synthesized nanocomposites were characterized using Powder X-Ray Diffraction (PXRD), FTIR, Scanning Electron Microscope (SEM) and Raman spectroscopic techniques. The antibacterial activity results of FGO showed significant activity against *S. mutans* (90-100%) and *Pseudomonas aeruginosa* (60-70%) compared to GO (40-50%), indicating the necessity of the functionalized ligand. Different parameters were optimized to achieve maximum activity. A possible explanation and mechanism are also proposed for the antibacterial activity.

[*] Corresponding Author's Email: amitdubeydr@manit.ac.in.

In: A Closer Look at Chemical Kinetics
Editor: Victor Martinez-Luaces
ISBN: 979-8-88697-484-3

Keywords: graphene oxide, antibacterial activity, sulfinic acid

Introduction

The search for new antibacterial drugs is imperative because of the increase in infectious diseases due to several pathogenic bacteria and the provocation of antibiotic protection [1, 2]. In recent years, pathogenic bacterial resistance from food-borne and community transmission has been growing continuously worldwide and has caused great concern regarding its impact on society [3, 4]. Several bacteria (gram-positive and gram-negative) such as *E. coli*, *S. aureus*, *S. mutans* and *Pseudomonas* are treated as emergent universal pathogens causing incurable life-threatening diseases [5, 6, 7].

In the current biotechnological era, nanoscale materials are emerging as innovative antimicrobial materials due to unique physicochemical characteristics (high surface area, better dispersion, etc.), and compared to the bulk materials [8, 9]. Different heterogeneous systems such as transition metal oxides, mixed oxides, bimetallic nanoparticles, silica materials and different plant extracts have been used as antibacterial agents for biomedical applications [2, 8, 10, 11, 12]. Several carbon nanomaterials have also been tested for drug delivery, molecular imagining and biosensing applications [13, 14, 15].

Over the last two to three decades, graphene oxide (GO) nanosheets have emerged as a potential material for many applications because of their size controllability, the flexibility of altering their properties by varying the oxidation state, and high solubility in solvents, thus enabling them to serve as better support for antibacterial activities [16, 17]. Biomedical activities such as cytocompatibility, toxicity and pharmacokinetics have been studied for graphene oxide materials. Polyethylene glycol (PEG) functionalized graphene oxide materials were reported for biomedical applications in pharmacokinetics and biodistribution for their toxicity in mice [18]. This study concluded that toxicity could not be mobilized with 20mg/kg dose over GO/PEG for several months.

In our continuous efforts to develop environment friendly heterogeneous nanocatalytic systems, different functionalized mesoporous silica and carbon nanocomposites were reported for catalysis, adsorption and antibacterial activities [19, 20, 21]. Several research reports have revealed that the majority of the carbon derivatives are toxic to bacterial cells, depending upon dose variation [22]. Additionally, antibacterial activity studies of functionalized graphene oxide, reduced graphene oxide and TiO_2 functionalized graphene oxide photocatalyst have been reported against different bacteria [2]. However, to the best of our knowledge, the literature has not so far reported the use of formamidine sulfinic acid (FSA) functionalized graphene oxide (FGO) against gram-positive and gram-negative bacteria.

Therefore, in order to showcase the synergistic effects of GO and the ligand (FSA), we report the synthesis and characterization of formamidine sulfinic acid (FSA) functionalized graphene oxide (FGO) for antibacterial activity of different gram-positive and gram-negative bacteria, especially on *Streptococcus mutans* (*S. mutans*) and *Pseudomonas* bacteria. The reason for choosing the ligand (FSA) is its reducing property as well as bleaching property [23]. Overall, four bacteria [*Staphylococcus aureus* (*S. aureus*), *Streptococcus mutans* (*S. mutans*)] [*Escherichia coli* (*E. coli*), *Pseudomonas bacteria*] were chosen for the study.

Materials and Methods

Synthesis of Graphene Oxide (GO) and Formamidine Sulfinic Acid Functionalized Graphene Oxide (FGO)

The chemicals and solvents purchased from Sigma Aldrich were used as such, without purification. GO was synthesized from graphite oxide using a modified advanced Hummer's method [24, 25, 26]. Typically, 3.0g of graphite oxide and 1.5g of $NaNO_3$ were mixed with 70ml of H_2SO_4 (concentrated, 98% pure) and stirred in an ice bath for half an hour by maintaining the temperature below 20°C. Thereafter, $KMnO_4$ (8g) was mixed carefully into the solution, and the solution kept in an ice bath followed by stirring for 3h at 25°C. 150ml of deionized water was added to the mixture and kept at 90°C for 15mins. When the color of the mixture started changing from brown to bright yellow, 500ml of deionized water followed by 8ml H_2O_2 were added to the solution. The resultant mixture after filtration was washed intensively with dilute HCl solution to remove impurities (metal ions and sulfate contamination) and then dried in vacuum at 40°C for 24h. The brown product obtained was dissipated in 600ml deionized water by ultrasonication. The solid GO was achieved by filtration and vacuum dried at 30°C for 36h [27].

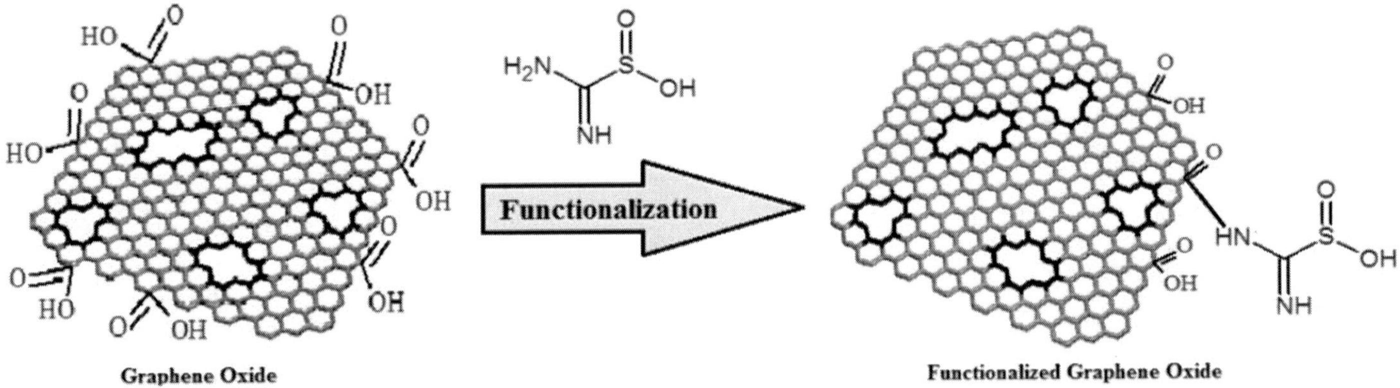

Scheme 1. Functionalization of graphene oxide.

Besides, for functionalization of GO, in a typical reaction 50mg/ml GO (20ml) aqueous dispersion was prepared through ultrasonication method. In addition, 0.5g of formamidine sulfinic acid (FSA) was added to the reaction under stirring and reflux conditions for 24h. Thus, the reaction mixture obtained was filtrated, washed with water and finally vacuum dried to obtain the brown, solid, functionalized graphene oxide (Scheme 1).

Physicochemical Characterization

PXRD

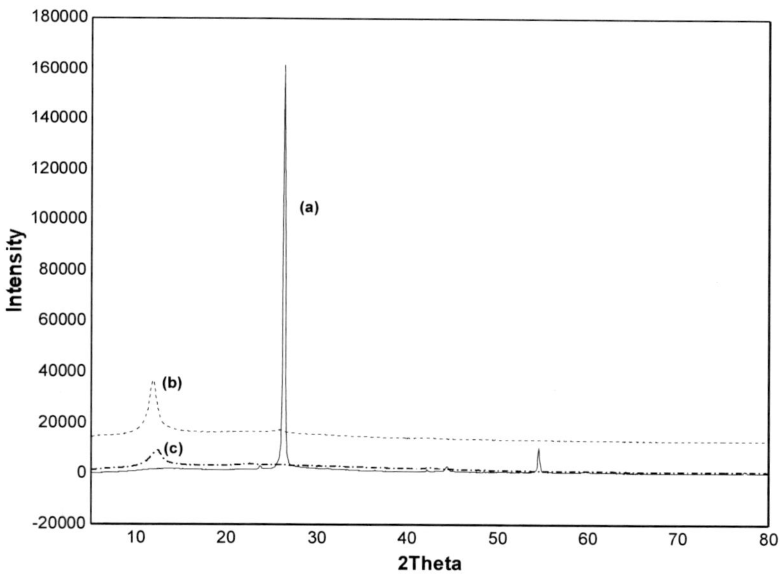

Figure 1. PXRD pattern of (a) Graphite, (b) Graphene oxide and (c) functionalized graphene oxide (FGO).

Figure 1 shows the PXRD patterns of graphene oxide (GO), graphite (G) and functionalized graphene oxide (FGO) nanocomposites. The graphite showed its characteristic diffraction peaks at 26.6, 44.7 and 54.6, analogous to (002), (101) and (004) peaks, respectively. The d-spacing (0.76nm) at 11.7° is due to the presence of oxygenated functional groups on both faces of GO and structural defects (roughness due to atomic scale). The structure of the functionalized graphene oxide sample was retained and confirmed by the corresponding diffraction peak of graphene oxide at 11.7° [11, 28]. The intensity of FGO decreased in comparison to the GO due to the reduction of the free surface functional groups, confirming the functionalization of FSA on GO.

Raman Spectroscopy

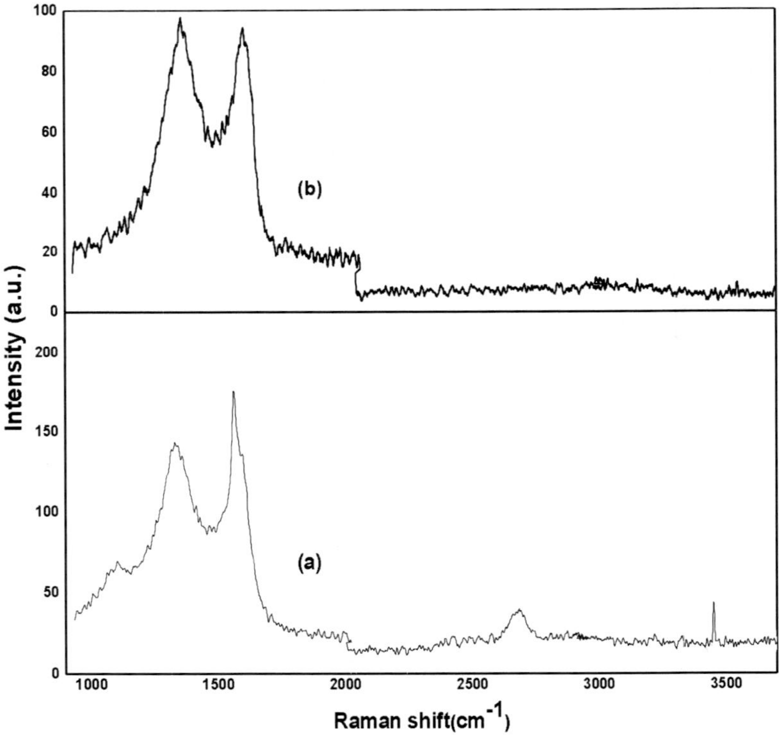

Figure 2. Raman spectra of (a) Graphite and (b) Graphene oxide.

FT-Raman was performed to see the structural changes during the synthesis of GO graphitic materials. Figure 2 shows that the peak position of the D band (breathing mode of k-photons (A_{1g} symmetry) at 1334 cm^{-1} and G-band (bond stretching motion (in-plane) of carbon- sp^2 hybridized atoms, E_{2g} symmetry) at 1564 cm^{-1} for both graphite and graphene oxide was retained, with higher peak intensity for GO material. In addition, the D band arose due to the synthesis of graphitic material involving a strong oxidation procedure.

The decrease in the intensity ratio I_g/I_D from 1.22 (for graphite) to 0.974 (for GO) showed the reduced size of in-plane sp^2 domains in comparison to a well-ordered graphite [29, 30].

FT-IR Studies

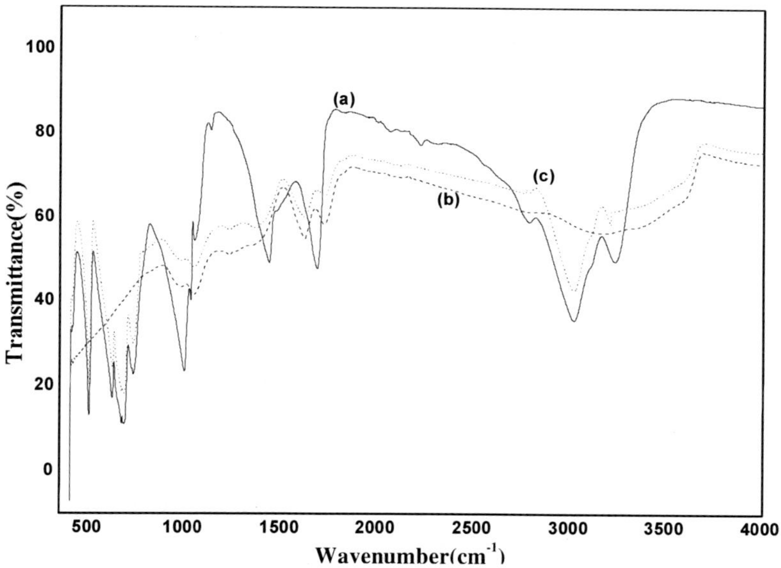

Figure 3. FT-IR spectra of (a) Formamidine sulfinic acid, (b) Graphene oxide and (c) functionalized graphene oxide (FGO).

The synthesis of GO and FGO was confirmed by FT-IR spectroscopy. The results in Figure 3 show that the peaks at 2800-3500, 1719, 1610 and 1052 cm^{-1} can be assigned due to the stretching vibrations of functional groups (O-H, C=C and C=O) in GO. Furthermore, the peak positions of the neat and functionalized ligand at 3233, 3025, 2780, 1687, 1437 and 980-700 cm^{-1} are due to –NH coupled, N-H symmetrical, S-H, C=N, C-N and S-O stretching vibrations and have not changed much after functionalization. An analysis of the results reveals that the broad peak due to amide N-H groups is merged with the broad peak of the O-H functionality and no bonding of the sulfonic acid groups has taken place. This is further substantiated by the presence of the amide carbonyl peak at 1670 cm^{-1}. All these results clearly confirm the functionalization of GO with the ligand [31].

Scanning Electron Microscopy and Atomic Force Microscopy

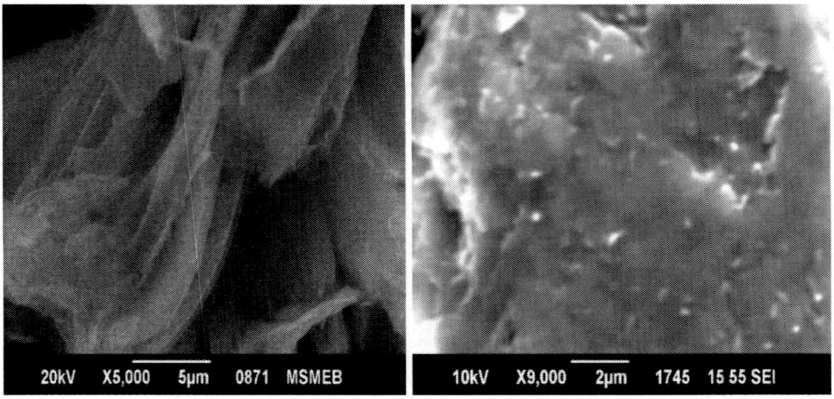

(a) (b)

Figure 4. SEM images of (a) Graphene oxide (GO) and (b) functionalized graphene oxide (FGO).

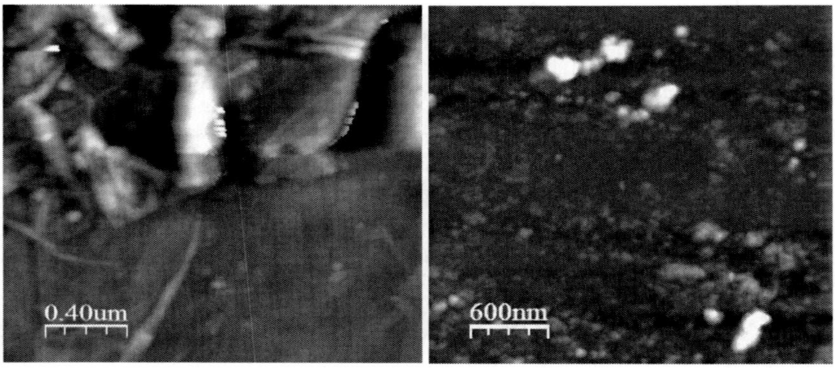

(a) (b)

Figure 5. AFM images of (a) Graphene oxide (GO) and (b) functionalized graphene oxide (FGO).

The SEM and Atomic force microscopy (AFM) images (Figure 4 and Figure 5) of GO can be seen as flake-like shapes with wrinkles on the surface and edges, whereas in FGO, particles of several nanometers are randomly loaded on the surface, indicating the favorable interaction between ligand and GO sheets [3].

Antibacterial Activity

Antibacterial activity of GO and formamidine sulfinic acid functionalized graphene oxide (FGO) was evaluated, against gram-positive bacteria: Staphylococcus aureus (*S. aureus*, ATCC-6538) and *Streptococcus mutans* (*S. mutans*, ATCC-25175), and gram-negative bacteria: *Escherichia coli* (*E. coli*, ATCC-2065) and *Pseudomonas aeruginosa* (ATCC-741), by finding minimum inhibitory concentration (MIC) using the micro broth dilution method and minimum bactericidal concentrations (MBC), by counting the number of colonies as the remaining bacteria in the culture.

The MIC can be explained as the minimum concentration of a drug at which no observable growth is noted in the broth medium, whereas MBC is the minimum concentration of a drug at which no growth of bacteria is noticed on the agar media plate.

Growth of Bacterial Strains

All bacteria were cultivated in Luria Bertani broth or LB media at 37°C and were grown up to mid exponential growth phase. Thereafter, cultures were centrifuged at 2500 rpm for 15 mins to separate the bacterial cells in pellet form, and cells were mopped up several times to take off residual macromolecules (growth medium constituents) with 9% isotonic aq. NaCl solution. Then, the bacterial cells were dispersed in 9% isotonic aq. NaCl solution and were diluted to get 10^6 to 10^7 colony forming unit (CFU)/ml cell concentrations.

Bacterial Vitality Test

All bacterial cells (suspended in isotonic NaCl solution) were treated with different concentrations of FGO and GO at 37°C under 200 rpm shaking speed for a day or 24h. The loss of vitality of the bacterial cells was estimated by optical density (OD) and by the colony counting method. The colonies formed on the agar media plate were identified and compared with those on control plates (without GO and FGO) to observe changes in the cell viability inhibition [32].

Results and Discussion

Antibacterial Activity of GO and FGO on Different Bacteria with Varying Concentrations

Initially, the antibacterial activity studies of GO and FGO on different bacteria were carried out at a fixed concentration of FGO (50ug/ml). Interestingly, very high activity of FGO was observed (60–100%) compared to GO (45–60%) for 4h under identical conditions, suggesting the necessity of the ligand FSA for enhancing the activity (Figure 6). This is due to the increase in the available active sites of the FGO to kill the bacteria more efficiently. Based on these results, *S. mutans* and *Pseudomonas* bacteria were selected for further optimization studies (Figure 7 and Figure 8).

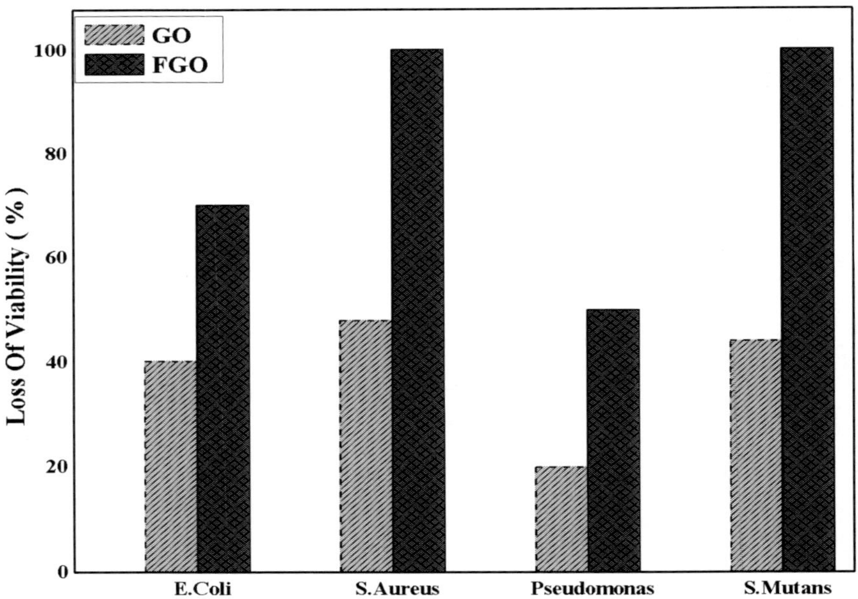

Figure 6. Antibacterial activity of different bacteria (viability measurements) over GO and FGO nanocomposites. *Conditions*: Concentration 50ug/ml, Time-4h at 37°C in 9% isotonic saline solution.

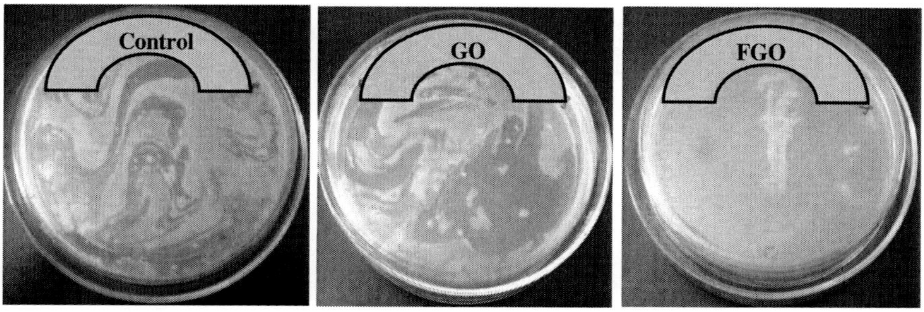

(a) (b) (c)

Figure 7. Bacterial colonies formed by *Pseudomonas* bacteria in agar medium (a) in the absence of GO/FGO (control experiment), (b) treated with GO (50ug/ml at 4h) and (c) treated with FGO (50ug/ml at 4h).

(a) (b) (c)

Figure 8. Bacterial colonies formed by *S. mutans* bacteria in agar medium (a) in the absence of GO/FGO (control experiment), (b) treated with GO (50ug/ml at 4h) and (c) treated with FGO (50ug/ml at 4h).

Effect of Concentration of FGO

The influence of concentration (10–50ug/ml) was evident, in that the growth of bacteria decreased with increase in the concentration of FGO (Figure 9 and Figure 10) and GO for all the bacteria (Table 1).

Maximum efficiency (100%) was achieved against gram-positive bacteria (*S. aureus* and *S. mutans*) and minimum (70%) for gram-negative bacteria (*Pseudomonas*), under identical experimental conditions. This can be due to the nature of the bacterial physiology, wherein only a thin layer of peptidoglycan surrounds the outer and inner cell walls of the gram-negative bacteria (*E. coli* and *Pseudomonas*), whereas the cell membrane of gram-positive bacteria (*S. aureus* and *S. mutans*) is mainly composed of thick peptidoglycan [33]. The loss of viability values (MIC and MBC) for all the bacteria shows that 40–45 ug/ml concentration of FGO is required for gram-positive and 50–80 ug/ml concentration for gram-negative bacteria (Table 2).

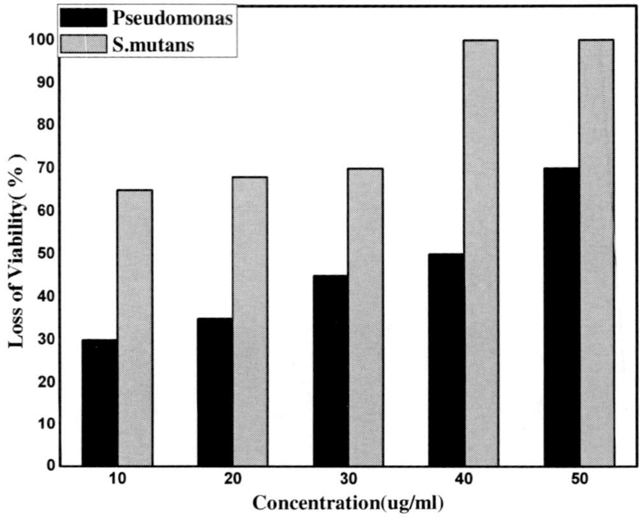

Figure 9. Variation of cell viability measurement of *Pseudomonas* and *S. mutans* bacteria over FGO nanocomposites with concentration. *Conditions*: Time- 4h at 37°C in 9% isotonic saline solution.

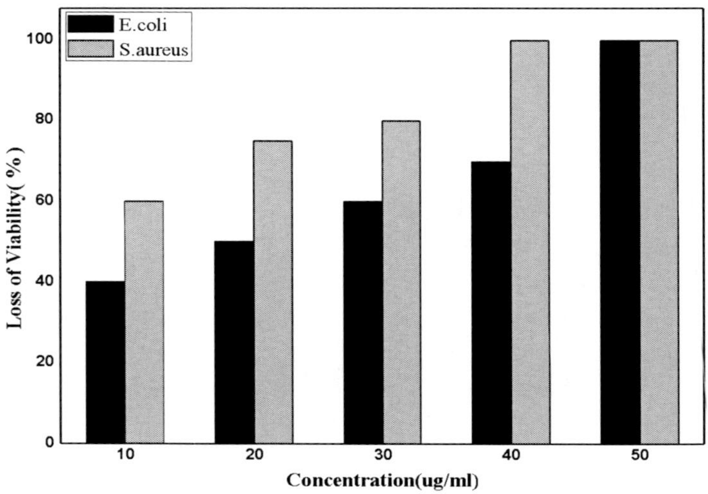

Figure 10. Variation of cell viability measurement of *E. coli* and *S. aureus* bacteria when incubated with different concentrations of FGO for 4h. *Conditions*: Time- 4h at 37°C in 9% isotonic saline solution.

Table 1. Influence of concentration of GO on different bacteria

Name of Bacteria	Concentration(ug/ml) of GO				
	10	20	30	40	50
E. coli	40	50	60	70	100
S. aureus	60	75	80	100	100
Pseudomonas	30	35	45	50	70
S. mutans	65	68	70	100	100

Growth Inhibition% = [1- (OD_{660} of Sample/OD_{660} of Control)* 100].
Reaction Conditions: Temp. - 37°C in 9% isotonic saline solution at 4h.

Table 2. Loss of viability values (MIC and MBC) for different bacteria with FGO

Name of Bacteria	MIC(ug/ml)	MBC(ug/ml)
S. aureus	40	43
S. mutans	40	45
E. coli	50	53
Pseudomonas	70	79

Effect of Contact Time with FGO

The effect of the contact time (Figure 11 and Figure 12) is reflected in the increased growth inhibition of bacteria (30–100%) with increase in the time of contact with FGO, in both cases.

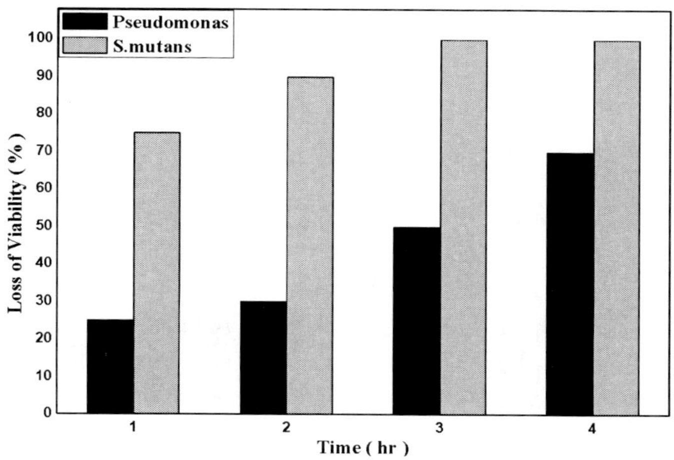

Figure 11. Variation of cell viability measurement of *Pseudomonas* and *S. mutans* bacteria over FGO nanocomposites with time. *Conditions*: Concentration-50ug/ml at 37°C in 9% isotonic saline solution.

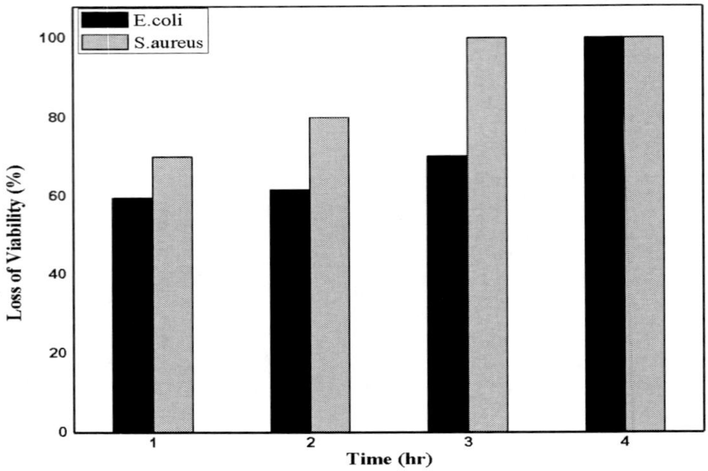

Figure 12. Variation of cell viability measurement of *E. coli* and *S. aureus* bacteria when incubated with FGO (50ug/ml) for given time intervals. *Conditions*: Concentration- 50ug/ml at 37˚C in 9% isotonic saline solution.

A careful analysis of the results shows that the growth of bacteria was repressed even at the beginning of the exposure. These results are advantageous over the previously reported results on graphene oxide against *Pseudomonas aeruginosa*, Ag/Zn functionalized graphene oxide activity against *E. coli* and *S. aureus* and Fe/Ag bimetallic supported graphene against *B. subtilis* and *S. aureus*, in terms of time and dose-dependent manner of cytotoxicity [11, 12].

Mechanism of Antibacterial Activity

Mechanistically, the amine group of FSA can easily interact with the proteins of the cell wall containing carboxylic acids and phosphorus-derived compounds in the cytoplasm, which provokes clogging of the respiratory system and cell division, resulting in loss of viability of the bacteria. In an aqueous medium, protonated amine groups of sulfinic acid have high binding affinity towards negatively charged bacterial cell membrane, and at the other end graphene oxides have a stronger adsorption capacity for bacteria because of the presence of lipopolysaccharide subunits in the outer membrane [34, 35]. Thus, the H-bonding facilitates the FGO nanocomposites adhering to the bacteria on its surface to increase the contact between the bacterial cells and

FGO. This eventually blocks the intake of nutrients by the bacteria, causing cell destruction [36]. Hypothetically, graphene oxide sheets are also considered to engender antimicrobial activity due to the membrane stress on the microbial cells caused by the sharp edges (Scheme 2).

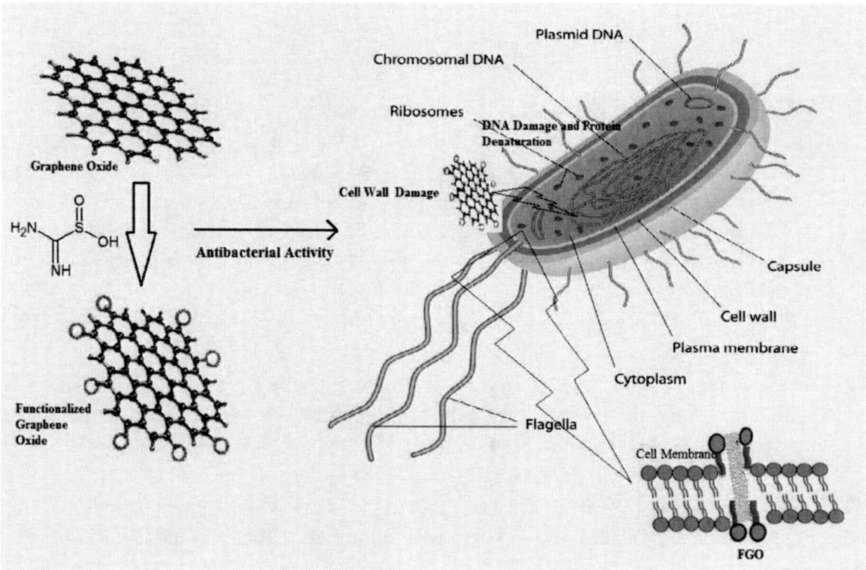

Scheme 2. Mechanism of antibacterial activity of FGO.

Thus, FGO nanocomposites act as inhibitors of bacterial growth and the increased antibacterial activity seems to be due to the synergistic effect between graphene oxide and the ligand.

Conclusion

Graphene oxide and functionalized graphene oxide were synthesized and characterized for antibacterial activity against gram-negative bacteria (*E. coli* and *Pseudomonas*) and gram-positive bacteria (*S. aureus* and *S. mutans*). The results of the broth microdilution method and the colony counting method showed very high antibacterial activity of the FGO compared to GO, substantiating the role of the functionalized ligand. The functionalization of GO with other ligands to deal with other bacteria also is currently underway.

Acknowledgments

The authors thank the Institute for their necessary help and support.

References

[1] Fan, Z., Po, K. H. L., Wong, K. K., Chen, S., Lau, S. P. Polyethylenimine-Modified Graphene Oxide as a Novel Antibacterial Agent and Its Synergistic Effect with Daptomycin for Methicillin-Resistant Staphylococcus Aureus. *ACS Applied Nano Materials* 2018, *1* (4), 1811–1818. https://doi.org/10.1021/acsanm. 8b00219.

[2] Wanag, A., Rokicka, P., Kusiak-Nejman, E., Kapica-Kozar, J., Wrobel, R. J., Markowska-Szczupak, A., Morawski, A. W. Antibacterial Properties of TiO2 Modified with Reduced Graphene Oxide. *Ecotoxicology and Environmental Safety* 2018, *147* (September 2017), 788–793. https://doi.org/10.1016/j.ecoenv. 2017. 09.039.

[3] Liu, S., Zeng, T. H., Hofmann, M., Burcombe, E., Wei, J., Jiang, R., Kong, J., Chen, Y. Antibacterial Activity of Graphite, Graphite Oxide, Graphene Oxide, and Reduced Graphene Oxide: Membrane and Oxidative Stress. *ACS Nano* 2011, *5* (9), 6971–6980. https://doi.org/10.1021/nn202451x.

[4] Ma, J., Zhang, J., Xiong, Z., Yong, Y., Zhao, X. S. Preparation, Characterization and Antibacterial Properties of Silver-Modified Graphene Oxide. *Journal of Materials Chemistry* 2011, *21* (10), 3350–3352. https://doi.org/10.1039/c0jm028 06a.

[5] Monte, J., Abreu, A. C., Borges, A., Simões, L. C., Simões, M. Antimicrobial Activity of Selected Phytochemicals against Escherichia Coli and Staphylococcus Aureus and Their Biofilms. *Pathogens* 2014, *3* (2), 473–498. https://doi.org/10.33 90/pathogens3020473.

[6] Wang, L., Hu, C., Shao, L. The Antimicrobial Activity of Nanoparticles: Present Situation and Prospects for the Future. *International Journal of Nanomedicine* 2017, *12*, 1227–1249. https://doi.org/10.2147/IJN.S121956.

[7] Hoerr, V., Zbytnuik, L., Leger, C., Tam, P. P. C., Kubes, P., Vogel, H. J. Gram-Negative and Gram-Positive Bacterial Infections Give Rise to a Different Metabolic Response in a Mouse Model. *Journal of Proteome Research* 2012, *11* (6), 3231–3245. https://doi.org/10.1021/pr201274r.

[8] Sharma, A., Robin Wilson, G., Dubey, A. Antibacterial ActiVIty of VInyl Imidazole(VI) Functionalized Silica Polymer Nanocomposites (SBA/VI) against Gram Negative and Gram Positive Bacteria. *New Journal of Chemistry* 2016, *40* (1), 764–769. https://doi.org/10.1039/c5nj01536g.

[9] Mishra, R. K., Ha, S. K., Verma, K., Tiwari, S. K. Recent Progress in Selected Bio-Nanomaterials and Their Engineering Applications: An Overview. *Journal of Science: Advanced Materials and Devices* 2018, *3* (3), 263–288. https://doi.org/ 10.1016/j.jsamd.2018.05.003.

[10] Vidic, J., Stankic, S., Haque, F., Ciric, D., Le Goffic, R., Vidy, A., Jupille, J., Delmas, B. Selective Antibacterial Effects of Mixed ZnMgO Nanoparticles. *Journal of Nanoparticle Research* 2013, *15* (5). https://doi.org/10.1007/s11051-013-1595-4.

[11] Gurunathan, S., Han, J. W., Abdal Dayem, A., Eppakayala, V., Kim, J. H. Oxidative Stress-Mediated Antibacterial Activity of Graphene Oxide and Reduced Graphene Oxide in *Pseudomonas* Aeruginosa. *International Journal of Nanomedicine* 2012, *7*, 5901–5914. https://doi.org/10.2147/IJN.S37397.

[12] Sandhya, P. K., Jose, J., Sreekala, M. S., Padmanabhan, M., Kalarikkal, N., Thomas, S. Reduced Graphene Oxide and ZnO Decorated Graphene for Biomedical Applications. *Ceramics International* 2018, *44* (13), 15092–15098. https://doi.org/10.1016/j.ceramint.2018.05.143.

[13] Dasari Shareena, T. P., McShan, D., Dasmahapatra, A. K., Tchounwou, P. B. A Review on Graphene-Based Nanomaterials in Biomedical Applications and Risks in Environment and Health. *Nano-Micro Letters* 2018, *10* (3). https://doi.org/10.1007/s40820-018-0206-4.

[14] Shi, J., Votruba, A. R., Farokhzad, O. C., Langer, R. Nanotechnology in Drug Delivery and Tissue Engineering: From Discovery to Applications. *Nano Letters* 2010, *10* (9), 3223–3230. https://doi.org/10.1021/nl102184c.

[15] Yun, Y. Carbon Nanomaterials: From Therapeutics to Regenerative Medicine. *Journal of Nanomedicine & Biotherapeutic Discovery* 2012, *02* (01), 1–8. https://doi.org/10.4172/2155-983x.1000104.

[16] Peng, W., Liu, S., Sun, H., Yao, Y., Zhi, L., Wang, S. Synthesis of Porous Reduced Graphene Oxide as Metal-Free Carbon for Adsorption and Catalytic Oxidation of Organics in Water. *Journal of Materials Chemistry A* 2013, *1* (19), 5854–5859. https://doi.org/10.1039/c3ta10592j.

[17] Qiu, J., Liu, L., Zhu, H., Liu, X. Combination Types between Graphene Oxide and Substrate Affect the Antibacterial Activity. *Bioactive Materials* 2018, *3* (3), 341–346. https://doi.org/10.1016/j.bioactmat.2018.05.001.

[18] Li, B., Zhang, X. Y., Yang, J. Z., Zhang, Y. J., Li, W. X., Fan, C. H., Huang, Q. Influence of Polyethylene Glycol Coating on Biodistribution and Toxicity of Nanoscale Graphene Oxide in Mice after Intravenous Injection. *International Journal of Nanomedicine* 2014, *9* (1), 4697–4707. https://doi.org/10.2147/IJN.S66591.

[19] Dubey, A. Synthesis and Catalytic Applications of CMK-LDH (Layered Double Hydroxides) Nanocomposite Materials. *Green Chemistry* 2007, *9* (5), 424–442. https://doi.org/10.1039/b616482j.

[20] Sachdev, D., Dubey, A. Mesoporous Silica/Polyphosphoric Acid (SBA-15/PPA) Nanocomposites for Acylation of Naphthalene. *Catalysis Communications* 2013, *39*, 39–43. https://doi.org/10.1016/j.catcom.2013.04.016.

[21] Verma, S., Dubey, A. Selective Debenzylation of 1-Benzyloxy-4-Methoxy Benzene over 5-Sulfosalicylic Acid-Functionalized Ordered Mesoporous Silica (SBA-15/SFSA) Catalyst. *Journal of Nanoparticle Research* 2017, *19* (12). https://doi.org/10.1007/s11051-017-4104-3.

[22] Mokkapati, V. R. S. S., Pandit, S., Kim, J., Martensson, A., Lovmar, M., Westerlund, F., Mijakovic, I. Bacterial Response to Graphene Oxide and Reduced Graphene

Oxide Integrated in Agar Plates. *Royal Society Open Science* 2018, *5* (11). https://doi.org/10.1098/rsos.181083.

[23] Umadevi, P., Deepti, K., Srinath, I., Vijayalakshmi, G., Tarakaramji, M. Synthesis and In-Vitro Antibacterial Activity of Some New Urea, Thiourea and Thiosemicarbazide Derivatives. *International Journal of Pharmacy and Pharmaceutical Sciences* 2012, *4* (SUPPL.3), 379–383.

[24] Shojaeenezhad, S. S., Farbod, M., Kazeminezhad, I. Effects of Initial Graphite Particle Size and Shape on Oxidation Time in Graphene Oxide Prepared by Hummers' Method. *Journal of Science: Advanced Materials and Devices* 2017, *2* (4), 470–475. https://doi.org/10.1016/j.jsamd.2017.09.003.

[25] Bourlinos, A. B., Gournis, D., Petridis, D., Szabó, T., Szeri, A., Dékány, I. Graphite Oxide: Chemical Reduction to Graphite and Surface Modification with Primary Aliphatic Amines and Amino Acids. *Langmuir* 2003, *19* (15), 6050–6055. https://doi.org/10.1021/la026525h.

[26] Naumov, A., Grote, F., Overgaard, M., Roth, A., Halbig, C. E., Nørgaard, K., Guldi, D. M., Eigler, S. Graphene Oxide: A One- versus Two-Component Material. *Journal of the American Chemical Society* 2016, *138* (36), 11445–11448. https://doi.org/10.1021/jacs.6b05928.

[27] Zaaba, N. I., Foo, K. L., Hashim, U., Tan, S. J., Liu, W. W., Voon, C. H. Synthesis of Graphene Oxide Using Modified Hummers Method: Solvent Influence. *Procedia Engineering* 2017, *184*, 469–477. https://doi.org/10.1016/j.proeng. 2017.04.118.

[28] Whitehead, K. A., Vaidya, M., Liauw, C. M., Brownson, D. A. C., Ramalingam, P., Kamieniak, J., Rowley-Neale, S. J., Tetlow, L. A., Wilson-Nieuwenhuis, J. S. T., Brown, D., McBain, A. J., Kulandaivel, J., Banks, C. E. Antimicrobial Activity of Graphene Oxide-Metal Hybrids. *International Biodeterioration and Biodegradation* 2017, *123*, 182–190. https://doi.org/10.1016/j.ibiod.2017.06.020.

[29] Hsiao, M. C., Liao, S. H., Yen, M. Y., Liu, P. I., Pu, N. W., Wang, C. A., Ma, C. C. M. Preparation of Covalently Functionalized Graphene Using Residual Oxygen-Containing Functional Groups. *ACS Applied Materials and Interfaces* 2010, *2* (11), 3092–3099. https://doi.org/10.1021/am100597d.

[30] Yang, S., Zhi, L., Tang, K., Feng, X., Maier, J., Müllen, K. Efficient Synthesis of Heteroatom (N or S)-Doped Graphene Based on Ultrathin Graphene Oxide-Porous Silica Sheets for Oxygen Reduction Reactions. *Advanced Functional Materials* 2012, *22* (17), 3634–3640. https://doi.org/10.1002/adfm.201200186.

[31] Kuila, T., Bose, S., Mishra, A. K., Khanra, P., Kim, N. H., Lee, J. H. Chemical Functionalization of Graphene and Its Applications. *Progress in Materials Science* 2012, *57* (7), 1061–1105. https://doi.org/10.1016/j.pmatsci.2012.03.002.

[32] Barer, M. R., Harwood, C. R. Barer. Bacterial Viability and Culturability. *Adv. Microb. Physiol.* 1999; 41:93-137. doi: 10.1016/s0065-2911(08)60166-6.

[33] Krishnamoorthy, K., Umasuthan, N., Mohan, R., Lee, J., Kim, S. J. Antibacterial Activity of Graphene Oxide Nanosheets. *Science of Advanced Materials* 2012, *4* (11), 1111–1117. https://doi.org/10.1166/sam.2012.1402.

[34] Xu, W. P., Zhang, L. C., Li, J. P., Lu, Y., Li, H. H., Ma, Y. N., Wang, W. Di, Yu, S. H. Facile Synthesis of Silver@graphene Oxide Nanocomposites and Their

Enhanced Antibacterial Properties. *Journal of Materials Chemistry* 2011, *21* (12), 4593–4597. https://doi.org/10.1039/c0jm03376f.

[35] Nanda, S. S., Yi, D. K., Kim, K. Study of Antibacterial Mechanism of Graphene Oxide Using Raman Spectroscopy. *Scientific Reports* 2016, *6* (March), 1–12. https://doi.org/10.1038/srep28443.

[36] Yang, K., Li, Y., Tan, X., Peng, R., Liu, Z. Behavior and Toxicity of Graphene and Its Functionalized Derivatives in Biological Systems. *Small* 2013, *9* (9–10), 1492–1503. https://doi.org/10.1002/smll.201201417.

Chapter 10

Electrodeposition of Zinc and Lead from Electric Arc Furnace Dust Dissolution: A Kinetic Study

Mayra Doldán[1], Martín Duarte[2], Gastón Pereira[2] and Mauricio Ohanian[1,*]

[1] IIQ, UDELAR, Montevideo, Uruguay

[2] Viento Sur Ingeniería [South Wind Engineering], Nueva Helvecia, Colonia, Uruguay

Abstract

The residue from the solids collection system at steelworks is known as electric arc furnace dust (EAFD) and it contains relevant amounts of iron, zinc and lead in oxide, silicate, and carbonate forms, among minor components such as chromium, tin, nickel, and cadmium. Thus, most countries classify this biproduct as a hazardous residue by environmental authorities. Its management has scientific and technical challenges that significantly could affect the economy of the steelworks process. The disposal of this residue at a reasonable cost involves thoroughly processing to recover valuable metals and separate or immobilize lead and cadmium. This process begins with a hydrometallurgical dissolution with the selective extraction of Zn, leaving the rest of the components in the solid. Zn has amphoteric properties but shares this characteristic with Pb, so alkaline extraction solubilizes both metals simultaneously, leaving iron compounds in an insoluble form.

After the dissolution step, two streams are obtained: a liquid and a solid. The liquid stream is a solution rich in Zn from which Zn will be

* Corresponding Author's Email: mohanian@fing.edu.uy.

In: A Closer Look at Chemical Kinetics
Editor: Victor Martinez-Luaces
ISBN: 979-8-88697-484-3

electrochemically recovered as a valuable product. The purity (Zn concentration) of the product obtained depends on the Zn:Pb ratio found in the dissolution step and also on electrochemical variables of the electrolytic stage (deposition potential, current density, electrode arrangement and fluid dynamic conditions of the system). The solid stream will be stabilized by incorporating it into an alkali-activated inorganic polymer (geopolymer) to obtain a product or residue that can be categorized as non-hazardous.

This chapter aims to analyze the electrochemical deposition step, evaluate the kinetics and characteristics of the product obtained, and use them to design the industrial process.

Keywords: Electric Arc Furnace Dust (EAFD), selective dissolution, electrodeposition, stabilization

Introduction

Electric Arc Furnace Dust (EAFD) is the common name for the particulate residue produced in any steel production route, generated in important amounts, and is expelled from the furnace through its exhaust gases, especially in processes that use steel scraps as raw materials. The resulting residue is retained in mechanical, electrostatic, and cyclonic filters, presenting relevant amounts of Fe and Zn in its composition. Those main components presence in the EAFD can be explained, besides Fe as the main component of steel, and Zn is derived from the dust due to the processing of galvanized scrap. Other minor elements are present, such as Pb, Cl, Si, Al, Mn, Cr, Mg, Ni, Cu and Cd, depending on the characteristics and origin of the processed scrap. The heavy metals present in the residue (lead and cadmium mostly) have prompted its classification as hazardous waste, thus effectively increasing the final disposition costs remarkably [1].

This issue is currently being investigated by many research groups, given the amount of residue generated and the additional costs that are associated. Treatment alternatives include pyrometallurgy and hydrometallurgical processes [2]. The pyrometallurgy process involves a reducing atmosphere at high temperatures. Even though the process is energetically intensive and only economically viable on a large production scale [3-5]. The hydrometallurgical process is more adaptable to a smaller EAFD production due to the lower initial investment and/or high cost energy conditions. The hydrometallurgical process includes serial stages with separation intermediate steps and recycling currents to minimize chemical costs and save energy costs. The first stage is selective dissolution, followed by the electrodeposition step. In this electroreduction, valuable metals are obtained, minimizing the global cost process.

The selective dissolution process aims to solubilize metals of commercial interest while leaving those that have no commercial value or are required for environmental reasons to remain in an insoluble form. At first, zinc was the main product to valorize from the EAFD processing. Metallic zinc could be used as anode material for cathodic protection [6] or recharge in hot-dip galvanizing baths [7]. The zinc compounds have amphoteric properties, and it is possible to achieve their dissolution in acidic or basic media. Due to that, alkaline conditions reach the dissolution of zinc while leaving iron, which is the other major compound, and other metals with similar behavior to remain insolubilized. Besides, the environmental stabilization of the solid residue (resultant from the dissolution step) is carried out in an alkaline medium [8-9]. Therefore, the selective dissolution in a basic medium presupposes reagents save in the final disposal stage. Regrettably, the lead shares the amphoteric behavior with zinc, so that the EAFD dissolution generates a zinc and lead alkaline solution without other metallic ions.

The electroreduction of lead or zinc in alkaline media is a well-known industrial process [10]. Nevertheless, the joining reduction generates a low purity (low price) zinc product. It is possible to perform a preliminary purification process, precipitating lead as sulfide [11].

Preliminary studies have shown that the lead sulfide precipitation is not completed in the present alkaline conditions. Besides, the zinc sulfide coprecipitates lowering the zinc recovery. On the other hand, the sulfide precipitation stage generates a new solid waste that must be stabilized and disposed of. Another possibility for the precipitation step is to develop a sequential electrodeposition, working in conditions that deplete the lead from the solution and prevent zinc deposition, and then move to a condition where the zinc can be obtained in high purity.

In this chapter, we present:

1) The theoretical and experimental modeling of the electrochemical electrodeposition of the EAFD basic leach, and
2) The development of a reactor prototype for the continuous recovery of zinc.

Electrochemistry Kinetics: Electrodics

The deposition reaction is electrochemical, therefore heterogeneous and at an electrified interface. The electrochemical reaction occurs at the metal/ electrolyte interface, but there are mass transport phenomena within the solution that must be considered in the global kinetics of the process. The passage of electrical current in the system generates net reactions of electrooxidation or electroreduction. The magnitude of the chemical effect is proportional to the circulating current, a relation found by Michael Faraday in the early 19th century. Therefore, in electrochemical systems, it is equivalent to express reaction rate and circulating current density (current per area). In the equilibrium state, the forward and reverse reaction rates are the same, being the circulating current densities the following [12]:

$$i_0 = i_c = i_a \quad (1)$$

where i_0 is the exchange current density and i_c and i_a, are the partial cathodic and anodic current densities.

If the system is not in equilibrium, the electrochemical potential (E) deviates from the value predicted by the Nernst equation. This difference from the equilibrium potential (E_{io}) is called system polarization or overpotential (η).

$$\eta = E - E_{i0} \quad (2)$$

In this condition, the net current density on the electrode surface is the result of the imbalance between the anodic and cathodic current densities:

$$i = i_c - i_a \quad (3)$$

Under non-equilibrium conditions, many phenomena occur on the electrode surface –ionic adsorption/desorption, solvatation, electron transfers, and the phenomena that occur within the electrolyte: transport of species, homogeneous chemical reactions, etc. On the other hand, due to the net passage of the current, ohmic effects are also generated. The mentioned phenomena generate irreversibility that is transferred to changes in the potential of the system, which is called overpotential (Equation 2). If one of these processes is slower than the others, it determines the global speed of the reaction and expresses that there is a control for the said stage. Under conditions of kinetic control by charge transfer on the electrode surface –for an elementary reaction– an expression can be deduced that relates the circulating current density to the overpotential. This relationship is called the Buttler-Volmer Equation (4):

$$i = i_0\left(\exp\left[-\frac{\beta F\eta}{RT}\right] - \exp\left[\frac{(1-\beta)F\eta}{RT}\right]\right) \tag{4}$$

β being the symmetry factor for an elemental reaction. If the reaction involves the transference of more than one electron, β is replaced by the factor α (charge transfer coefficient), which is a linear combination of symmetry factors of elementary steps.

The Buttler-Volmer equation (4) has two approximations. At overpotentials close to zero, the dependence between the current density and the overpotential is linear:

$$i = i_0\frac{\alpha F\eta}{RT} \tag{5}$$

While at high overpotentials, Equation (4) takes a logarithmic form:

$$\ln i = \ln i_0 + \frac{\alpha F\eta}{RT} \tag{6}$$

This well-known equation was found by Julius Tafel for the evolution of hydrogen, and it is valid for any system that has activation control.

The Buttler-Volmer expression is applicable to electrochemical phenomena controlled by charge transfer; in the case of mass transfer control, the charge transfer reaction can be considered as an instantaneous process. The control by mass transfer and the consumption of species on the surface of the electrode generate a difference in concentrations between the bulk of the solution and the electrochemical interface. Species transport phenomena are due to convection (caused by mechanical agitation or thermal gradients), migration (movement of charged species due to electrical potential gradients) and diffusion (transport of species due to chemical potential gradients). Considering that the electrochemical reduction generates a consumption of the electroactive species, a decrease in the concentration on the surface of the electrode occurs. The chemical gradient causes the appearance of the diffusional process. Considering a one-dimensional phenomenon, the flux J (in moles per unit area and per unit time) of species k is expressed:

$$J = -D_k \left(\frac{\partial C_k}{\partial x} \right) \tag{7}$$

The proportionality constant D_k is called the diffusion coefficient and C_k is the concentration of the k species. Equation (7) is called Fick's first (one-dimensional) law. It is an empirical relationship that was found by Adolf Fick in 1855 [13].

If the determining stage of the process is the diffusion of species, it is considered that the electroactive species reacts instantaneously when it reaches the surface of the electrode. Under these conditions, the electric current density is limited by the flow of species that reach the surface and by applying Faraday's relation:

$$i = -nFJ\big|_{x=0} \tag{8}$$

F being the Faraday constant, taking the local flux at the electrode surface, $x = 0$.

For the non-steady states, the change in the concentration profile with time can be found by applying Fick's second law (which is deduced from establishing a mass balance in a control volume within the electrolyte). Expressed in one-dimensional form, it is written as:

$$\frac{\partial C_k}{\partial t} = D_k \frac{\partial^2 C_k}{\partial x^2} \tag{9}$$

The electrode potential difference for a system in equilibrium is related to the surface activities from the Nernst equation. The potential difference of the electrode at a given current density with respect to the equilibrium situation, is linked to the ratio of the concentration in the solution and the surface of the electrode:

$$E - E_{eq} = \eta_{dif} = \frac{RT}{nF}\ln\left(\frac{C_k^{x=0}}{C_k^{Bulk}}\right) \tag{10}$$

η_{dif} being the diffusion overpotential.

It can be considered that at distances greater than or equal to a value called the thickness of the Nernst diffusion layer (δ), the concentration is equal to the bulk.

If the reaction is:

$$A + e^- \rightarrow B \tag{11}$$

Equation (9) can be solved by applying as an initial condition that the reagent concentration is uniform throughout the electrolyte and that the product concentration is equal to zero: $C_{A,t=0} = C_A^{Bulk}$ and $C_{B,t=0} = 0$.

As boundary conditions, we can consider that the concentration of reactant and product are constant on the electrode surface:

$$C_{A,t>0}^{x=0} = C_A^{x=0} \text{ and } C_{B,t>0}^{x=0} = C_B^{x=0} .$$

The linearized concentration profile for locations near the electrode surface can be written as:

$$C_A(x) = C_A^{x=0} + \left(\frac{C_A^{Bulk} - C_A^{x=0}}{\delta}\right)x \tag{12}$$

Consequently, the concentration gradient is equal to:

$$\frac{dC_A(x)}{dx} = \frac{C_A^{Bulk} - C_A^{x=0}}{\delta} \tag{13}$$

It presents a maximum value for the surface concentration equal to zero, for which there will be a maximum current density. This current density is called the diffusional limit current density, i_d.

$$i_d = nFD_A \frac{C_A^{Bulk}}{\delta} \tag{14}$$

Combining (13) and (14):

$$\frac{C_A^{x=0}}{C_A^{Bulk}} = \left(1 - \frac{i}{i_d}\right) \tag{15}$$

If we substitute expression (15) in (10), we can relate the diffusion overpotential to the relationship between the current density and the limiting current density (experimentally determinable parameters):

$$\eta_{dif} = \frac{RT}{nF} ln\left(1 - \frac{i}{i_d}\right) \tag{16}$$

At current limit values, the species are reduced as soon as they reach the electrode surface. Under this condition, there is no reagent concentration on the electrode's surface. If using an external source, the system is forced to a current above the limit current, the electrode potential will change with the appearance of other reduction reactions on its surface. Very rough or dusty deposits are obtained when the solution in the vicinity of the cathode is depleted with respect to metal cations.

The structure of the deposit depends on whether the characteristic height of the surface roughness is smaller or larger than the thickness of the diffusion layer. For a height smaller than the diffusion layer (micro profile) the peaks are favored concerning diffusion because the free cross-section for diffusion increases with an increasing distance from asperity and more metal is deposited on the peaks than in the valleys. In contrast, the generation of macro profiles –with dimensions greater than the thickness of the diffusion layer– increases the possibility of deposition both at the peak and in the valleys of the roughness profile and a greater tendency to the formation of compact deposits. Therefore, to promote the appearance of powdery deposits, it is necessary to work under mass transport control and agitation conditions to generate micro profile conditions on the cathode surface [14].

Materials and Methods

The EAFD used in this study was obtained by sampling and mixing different extracted samples for six months. The chemical composition and morphology were determined by scanning electron microscopy (SEM) equipped with an EDAX microprobe. Before the morphological observation, a gold coating must be performed to improve the quality of the images (sputtering).

The samples were subjected to an X-ray diffraction test for crystallographic characterization. The equipment used was a powder diffractometer Rigaku Miniflex, CuKalpha, 40 KV, 15 mA, 0.02 deg step, 20 deg/min.

A preliminary dissolution stage was necessary for the electrochemical tests to obtain the dissolution rich in zinc and lead in a basic medium. The EAFD was subjected to a basic dissolution process in a NaOH/powder ratio of 3.5 g/g, the H2O/EAFD ratio was 10 mL/g. The mixture was heated at 80 °C and stirred for 30 minutes. Subsequently, the supernatant was separated from the sludge by centrifugation.

The zinc and lead solutions (Aldrich Reagent) were prepared in 0.01 M for lead and 0.44 M for zinc in alkaline media.

The electrochemical experiences was performed in a 1000 mL glass (Radiometer® CNC). The electrolytes used were the solutions of zinc, lead, and steel dust solutions. The studies were carried out at 20ºC without agitation. The cell arrangement consists of three electrodes: a working electrode (stainless steel AISI 304), a platinized platinum counter-electrode, and a saturated $Hg/HgSO_4.K_2SO_4$ reference electrode (all the potentials in the rest of this work are referred to this reference electrode), + 0.710 V vs. normal hydrogen electrode (NHE). The working electrode surface was polished to 220 grit before each experience. A Gamry Instrument Reference 600+ equipment is used as an interface and potential control. The electrochemical tests consist of lineal voltammetry (scanning speed is 5 mVs-1, Ecorr +0.25V to Ecorr-3V) and potentiostatic pulses (chronoamperometric, 600 s) experiences.

After each chronoamperometric experiment, the deposits were characterized by X-ray fluorescence (XRF) Thermo Scientific Niton X-ray fluorescence, Niton XL2 950 GOLDD).

Results and Discussion

Characterization of Steel Mill Waste

From the SEM observation, it can be seen that the particle sizes present in the residue, are grouped into two clusters, one with average sizes between 1 and 2 μm and the other with approximately 20 μm, see Figure 1. EDAX analysis of the different particle sizes show similarity in their elemental composition with mass concentration in: oxygen (28.2%), iron (16.5%), zinc (36.0%) and lead (0.82%).

Figure 1. EAFD SEM image.

The spectrum from the XRD analysis is shown in Figure 2. The crystalline compounds present in the EAFD correspond mainly to magnetite Fe_3O_4, zinc oxide (ZnO), and cristobalite (SiO_2).

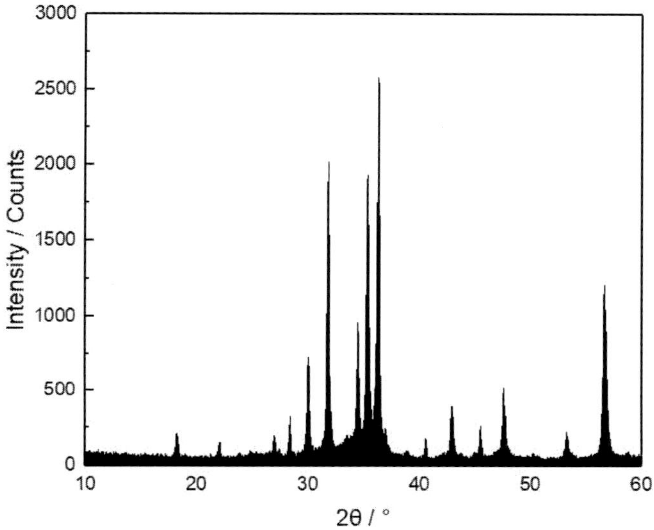

Figure 2. XRD spectrum obtained from the EAFD residue.

Electrochemical Evaluation

The voltammetric responses for lead, zinc, and EAFD dissolutions are shown in Figure 3.

A steady growth can be seen in the lead standard when the potential decreases. The electrodeposition of metallic lead overlaps with the hydrogen evolution reaction (HER) [15-16]:

$$HPbO_2^- + H_2O + 2e^- \rightarrow Pb + 3OH^- \qquad E=-1.187\ V \qquad (17)$$

$$2H_2O + 2e^- \rightarrow H_2 + 2OH^- \qquad E=-1.667V \qquad (18)$$

For the standard zinc solution, a peak towards -1.6 is distinguished and attributed to the evolution of hydrogen and a subsequent increase due to the electroreduction of zinc [18]:

$$Zn(OH)_4^{2-} + 2e^- \rightarrow Zn + 4OH^- \qquad E=-1.849\ V \qquad (19)$$

For the EAFD dissolution, there is a visible decrease in the corrosion potential towards more negative values with respect to the pure metal solutions. Current jumps at potentials of -1.5, -1.6 and -1.8 V could be appreciated and be attributed to the reactions (17), (18) and (19) respectively. In the voltammetry the current values are of similar magnitude; so, it is expected that once the zinc deposition potential is reached, the reduction kinetics for the three reactions is similar and in the order of 0.1 A/cm2. The current density limit reaches values on 1 A/cm2.

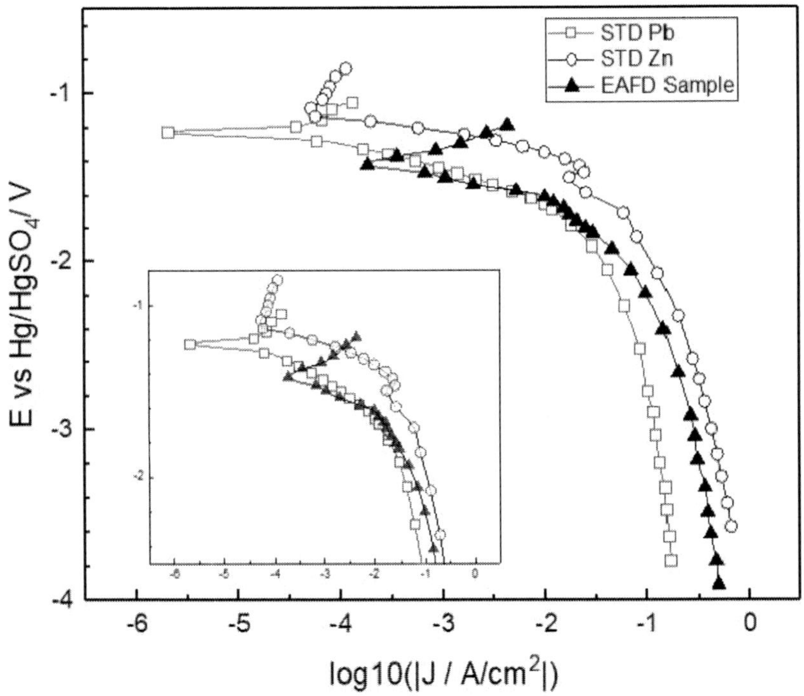

Figure 3. Voltammetric curves for lead, zinc and EAFD dissolutions.
In the enlargement, it is the zone near the corrosion potential.

The potentiostatic steps curves (chronoamperometric curves) are shown in Figure 4.

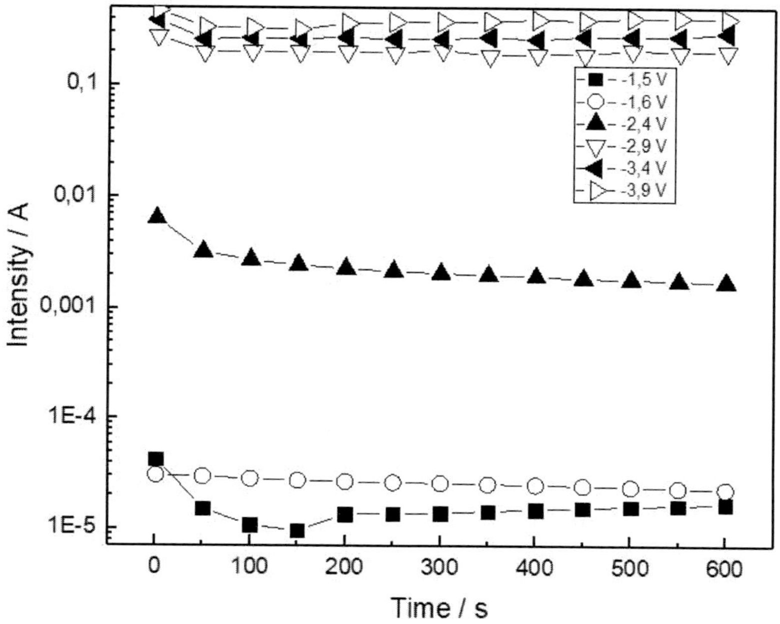

Figure 4. Chronometric curves of EAFD dissolution.

The data of the chronoamperometric curves are adjusted according to the Cottrell equation [19]. For the development of Cottrell equation, mass transfer control conditions, a single species is electroreduced, a flat surface electrode, homogeneity of the solution, and semi-infinite diffusion are assumed:

$$i(t) = \frac{nFAD^{\frac{1}{2}}C^{Bulk}}{\pi^{\frac{1}{2}}t^{\frac{1}{2}}} = K\, t^{-\frac{1}{2}} \tag{20}$$

A being the electrode area, *D* the electroactive species diffusivity, and C^{Bulk} the bulk concentration.

If the power of the time parameter is left as a degree of freedom, the equation (20) takes an allometric format and can be fitted to the experimental chronoamperometric curves obtained.

$$i(t) = at^{b} \tag{21}$$

Table 1 shows the adjustment results for the parameters *a* and *b*:

Table 1. Allometric fit parameters for the applied potentials

E (V)	a	B
-1.5	1.82E-5	-0.041
-1.6	3.65E-5	-0.062
-2.4	7.17E-3	-0.21
-2.9	2.11E-1	-0.0094
-3.1	2.41E-1	-0.016
-3.4	2.48E-1	0.017
-3.9	2.71E-1	0.055

The allometric parameter *a* increases as the imposed potential decreases, revealing an increase in the global reduction kinetics. However, the potential factor *b* shows a marked difference from the theoretical Cottrell value. This behavior can be explained by several reasons:

a) The non-faradaic components of the electrochemical double layer [20] released during the imposition of the jump in the potential have been neglected. The double layer can be modeled by resistive capacitive electrical components, whereby an exponential decay behavior is superimposed.
b) The presence of three existing cathodic reactions on the same surface: the potentials of the electrodepositions overlap with the evolution of hydrogen, so it cannot be considered. Once zinc deposition begins, the rest of the kinetics is dominated by mass transfer [12].
c) The reaction on a flat electrode represents only the initial condition of electrodeposition. The deposit growth is only flat and compact at low reaction rates, while at low potentials and high deposit rates, the morphology is dendritic (see Figure 5).

During the growth of the metallic deposit, the macroscopic and microscopic areas are different, in fact, the microscopic surface is irregular and partially active. On the other hand, area *A* is included in the Cottrell equation regardless of whether it responds to a macro or micro profile compared with the thickness of the diffusion boundary layer. This phenomenon has an influence on the dependence of the current with time in the chronoamperometric test.

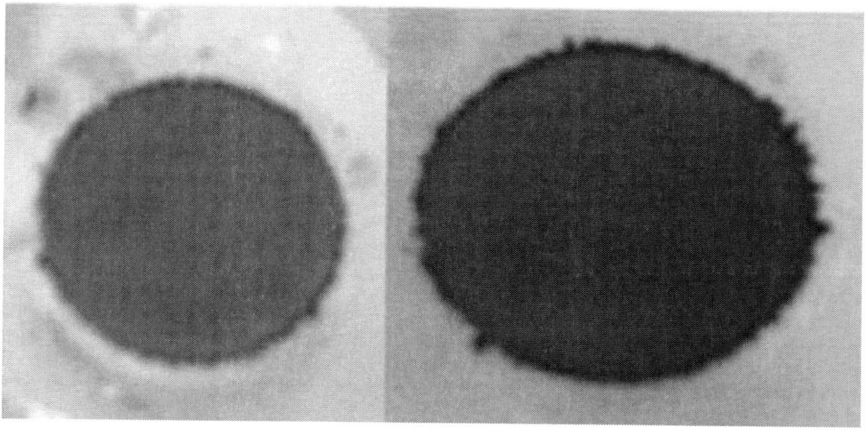

Figure 5. Photographs of deposits at -1.5V (left) and -2.5V (right). Note the dendritic growth on the edge of the electrode.

This event is studied by Pajkossy, who proposes that three-time intervals can be distinguished in the chronoamperometric test: 1) initial times, 2) final times and, 3) intermediate times. At initial and final times, the dependence corresponds to the macroscopic and microscopic areas respectively, with a dependence like the Cottrell form. At intermediate times, there is a dependency on the surface characteristic fractal length (Df) [21]. The temporal dependency would be reliant on the exponent, α, being alpha

$$\alpha = \frac{D_f - 1}{2} \tag{22}$$

Nevertheless, the fits for initial, middle, and final time intervals, do not reveal clear tendencies in the experimental chronoamperometric curves. In most cases, the b parameter is between -0.1 and -0.3 values in the initial intervals, takes values over -0.1 in middle times and reach positive values (near 0.2) in the last intervals. The observed behavior shows that there is more than one cause that explains the difference from the theoretical Cottrell relationship [22-23].

For the chronoamperometric tests developed, XRF determination of the composition of the deposit are carried out once the experience has ended. The results are presented in Figure 6.

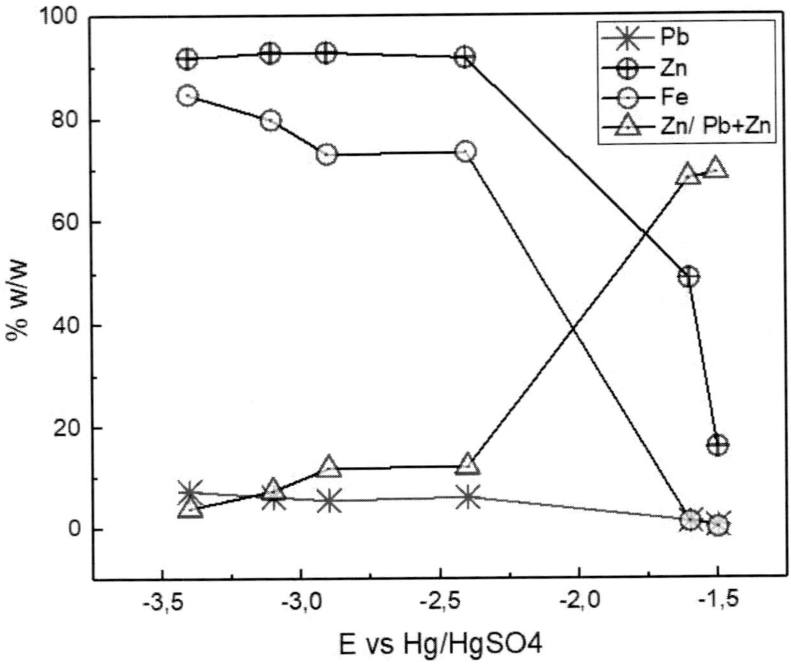

Figure 6. Chemical composition vs applied potential of the deposits obtained as a result of the amperometric tests.

In Figure 6, the mass percentage for the two electrodeposited elements (lead and zinc) and iron –the main component of the base electrode stainless steel– can be observed. As the test potential decreases, the determined iron concentration also decreases due to the increment of the thickness electrodeposition. The deposit is rich in lead at high potentials; below -2 V the zinc layer constitutes approximately 90% of the deposit. This fact has a key technological consequence in the implementation of the system on an industrial scale as it is possible to exhaust the lead solution in a first stage imposing a high potential, and move on to a second stage, where a zinc deposit is obtained. In this way, high purity deposits with a higher commercial value can be obtained. However, the control of potentiostatic processes is more difficult (due to the potential control system [24]) than a current control system (galvanostatic, see Figure 7). With a galvanostatic system, a current that prioritizes the deposition efficiency (deposited metal/consumed current) is imposed, obtaining an alloyed metal product (lead/zinc) with less commercial value than pure products. The decision is financial-economical.

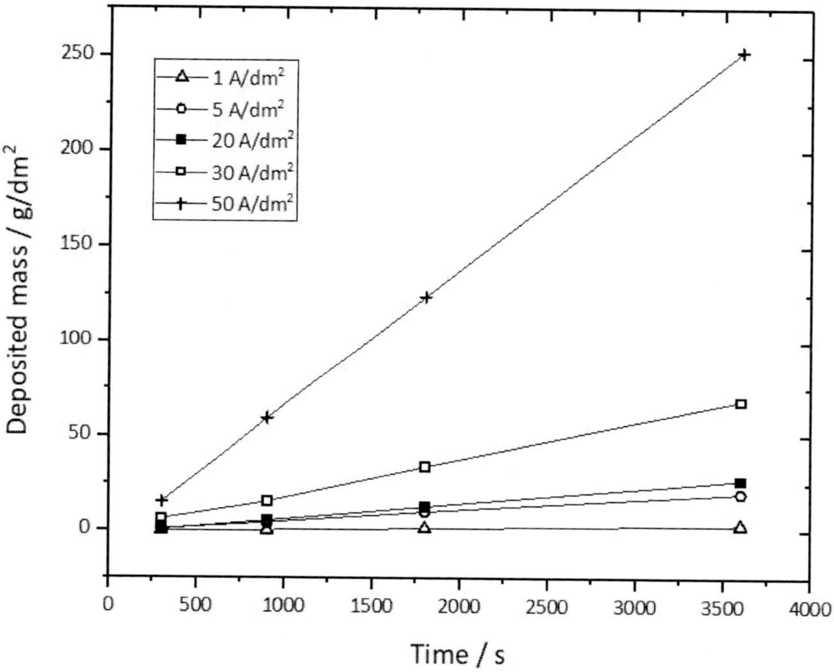

Figure 7. Profile of mass deposited vs. time at different current densities.

Pilot Scale Reactor Electrolyzer

The development of a rotary electrolyzer on a pilot scale is proposed in order to have a basis for industrial scaling. At this stage, it should be considered the fluid dynamic flow model of the reactor, the nature of the working electrode, the method of current supply for individual cells, and the arrangement of electrolyte flow through the cell. Likewise, the nature of the products obtained must be considered and a metallic solid is obtained in the cathode. On the other hand, the hydrogen gas evolution should be seen as a secondary cathodic reaction. At the anode (if it has an inert nature) there will be a release of gaseous oxygen.

The choice of a rotating cylindrical geometry is due to the possibility of electrodepositing with continuous product removal while generating the necessary agitation to improve mass transport. Other types of configurations or geometries are simpler to implement regarding power supply, sealing and product inputs/outputs; however, they do not solve the continuous operation [25].

It is proposed the use of rotating cylindrical electrode (RCE) geometry with continuous removal of the electrodeposited metal. The previously mentioned configuration can be referenced for small-scale electro-organic synthesis and the recovery of metals from effluents [26, 27]. The bibliography proposes rotation schemes with the vertical and horizontal axis without specifying construction details. It is pointed out that most of the technical developments are protected by industrial patents. Hence, it can only be accessed to partial schemes and theoretical studies with emphasis on fluid dynamics, mass transfer, and electrochemical considerations.

Most of the RCE designs are applied in metallic electrodeposition work with fairly high metal concentrations and at moderate current densities and with relatively low peripheral velocities (laminar flow conditions and low mass transfer). Such conditions produce sticky electrodeposits. These conditions are contrasted with reactors, such as the Eco-Cell, where high peripheral speed, turbulent flow, high mass transfer, and deposition of rough dust and conditions that facilitate the treatment of low metal concentration solutions and allow recovery at established concentrations less than 10%. This reactor is capable of continuously recovering metals, from a dilute solution, as metal powder. It allows improving mass transfer. It is used in a continuous flow, in a divided cell, the cathodic electrolyte taking the detached dust and hydrogen gas to a three-phase separator system. The anode is separated from the cathode by a semi-permeable membrane. The control of the system is – from the electrochemical viewpoint, potentiostatic, which allows selectivity in the electroseparation. The cell is designed to operate under turbulent conditions. The conversion of the system –in one step– can be improved by increasing the mass transport coefficient (by increasing the rotation speed, or by a further increase in the electrode surface area). However, there are practical limitations in cost, space, and engineering to increase the conversion of a single stage, so the use of reactors fed in hydraulic series is proposed. The Eco-Cascade cell uses this principle by assembling the entire reactor on a single shaft to minimize equipment costs and power consumption [28].

Based on the electrochemical, fluid dynamics, and transport mass system characteristics, we build a pilot scale reactor. It has a rotating cylindrical cathode and a static anode. The EAFD solution is fed in the batch. The cathode has a scraper that continuously removes the metallic deposit. The positive pole of the direct current source is connected directly to the anode, whereas the negative pole is connected to the cathode by a carbon terminal. A variable speed motor turns the cathode shaft (Figure 8).

Figure 8. CAD design and photograph of the prototype electrolyzer.

Additionally, there have been experiences with the prototype proposed with the galvanostatic mode. During the experiences, some difficulties in the product separation were identified, due to mechanical causes which were solved afterwards. At low current densities, the metallic deposit is hard and adherent. However, at high current densities the deposit is porous and can be removed easily but at the expense of losing current efficiency. At this time, the prototype is being designed on an industrial scale.

Conclusions

- The elemental crystallographic composition, chemical and electrochemical behavior of the EAFD produced in a local metallurgy have been characterized.
- The residue has been treated by hydrometallurgical means to reduce and stabilize the residues produced.
- The kinetic theory of electrodeposition has been analyzed considering the chemical, electrochemical, and mass transfer conditions that allow obtaining a metallic product that can be recovered from the EAFD leachate.
- Metallic zinc has been recovered, which valorizes part of the EAFD waste. The main contaminant in the zinc obtained is lead. The degree of purity of zinc depends on the chemical or electrochemical process,

which changes the cost of processing and the economic value of the product obtained.

- A pilot-scale electrolyzer has been developed, allowing the obtention of zinc in a continuous process. Thus, electrochemical efficiency and construction improvements permit advance in the project's chemical design at an industrial scale.

References

[1] Walburga Keglevich de Buzin, P. J.; Heck, N. C.; Faria Vilela, A. C.; *J. Mater. Res. Technol.*, 2017, 194-202.

[2] Rossini, G.; Moura Bernardes, A. J. *J.; Hazard. Mater.*, 2006, 210-216.

[3] Mackey, P.J.; Balfour, G. C. *J. Met.*, 1978, 36-43.

[4] Palimaka, P.; Pietrzyk, S.; Stepien´, M.; Ciecko, K.; Nejman, I.; *J. Met.*, 2018, 8, 547.

[5] Grillo, F. F.; Coleti, J. L.; Espinosa, D. C.; Oliveira, J. R.; Tenório, J. A. S., *Mater. Trans.*, 2014, 351-356.

[6] Shreir L. L., Jarman R. A., Burstein G. T.; *Corrosion Control*; ISBN 0 7506 1077 8; Butterworth Heinemann: Oxford, GB,1994; pp. 10.1-10.121.

[7] Roberge, P. R.; *Handbook of Corrosion Engineering*; ISBN 0-07-076516-2; McGraw-Hill: New York, 2000; pp. 863-920.

[8] Tran Huyen, V.; Nadarajah, G., *Adv. Concr. Technol.* , 2018, 124-135.

[9] Provis, J.; Van Deventer, J.; *Alkali Activated Materials*; ISBN 978-94-007-7672-2; Springer: Dordrecht Heidelberg, New York, London, 2014; Vol.13, pp. 93-157.

[10] Winand, R.; *Modern Electroplating, Schlesinger;* ISBN 978-0-470-16778-6, John Wyley and Sons: New Jersey, 2010; pp. 285-308.

[11] Varga, T.; Bokányi, L.; Török, T. I.; *Int. J. Miner. Metall. Mater.*, 2016, 2-121.

[12] Bockris, J. O´M.; Reddy, A. K. N.; Gamboa-Aldeco M. *Modern Electrochemistry Second Edition Fundamentals of Electrodics*; ISBN 0-306-46166-8, Kluwer Acedemics Publishers, New York, 2002; pp. 1042-1073.

[13] Bird R.B.; Stewart W. E.; Lightfoot E. N. *Transport Phenomena*; Card Number 60-11717; John Wyley Sons: Singapore,1960; pp. 495-518.

[14] Ibl N. *; Surf. Coat. Technol*, 1980,10, 81–104.

[15] Zheng, Y.; Jiao, Y.; Vasileff, A. and Qiao, S.(2017) *Hydrogen Evolution Reaction in Alkaline Solution: From Theory, Single Crystal Models, to Practical Electrocatalysts*. Wyley-VCH 10.1002/anie.201710556.

[16] Wong, S.M; Abrantes, L.M.; *Electrochim. Acta.,* 2005, 51, 619-626.

[17] Vanysek, P. in CRC Handbook of Chemistry and Physics, David. R. Lide Ed., pp. 8.20-8.29, *CRC Handbook of Chemistry and Physics*, Internet Version 2007, (87th Edition) Taylor and Francis; http:/www.hbcpnetbase.com.

[18] Ohba, M.; Scarazzato. T.; Espinosa, D. C. R.; Panossian, Z.; *Electrochim. Acta.* 2019, 309, 86-103.

[19] Bard, A. J.; Faulkner, L. R.; *Electrochemical Methods Fundamentals and Applications*; ISBN 0-471-05542-5; Jhon Wyley & Sons: New York, 1980; pp. 142-148.
[20] Wendt, H.; Kreysa, G.; *Electrochemical Engineering Science and Technology in Chemical and Other Industries*; ISBN 978-3-662-03851-2; Springer-Verlag: Berlin Heidelberg; 1999; pp. 39-80.
[21] Pajkossy,T.; *Encyclopedia of Interfacial Chemistry Surface Science and Electrochemistry*, ISBN 978-0-12-809739-7; Elsevier: MA USA, 2018; pp. 121-124.
[22] Diaz, V.; Ohanian, M.; Zinola, C. F.*; Int. J. Hydrog*, 2010, 10539-10546.
[23] Pajkossy, T.; Nyikos, L.; *Electrochim. Acta*. 1989, 34, 34-171.
[24] Carrillo-Abad, J.; García-Gabaldón, M. Peréz-Herranz, *V.; Int. J. Electrochem. Sci.* 2012, 7, 5442-5456.
[25] Youcai, Z. C. *Pollution Control and Resource Reuse for Alkaline Hydrometallurgy of Amphoteric Metal Hazardous Wastes*; ISBN 978-3-319-55157-9; Springer International Publishing: China, 2017; Vol 18, pp. 171-259.
[26] Goodridge, F.; Scott, K.; *Electrochemical Process Engineering A Guide to the Design of Electrolytic Plant*; ISBN 978-1-4899-0226-9; Springer Science Business Media: New York, 1995; pp. 177-240.
[27] Gardner, N. A.; Walsh, F. C.; *The Design and Application of Rotating Cylinder Electrode Technology to Continuous Production of Metal en Electrochemical Cell Design*. Ed. Ralph E. White; ISBN-13: 978-1-4613-2795-0; Plenum Press: New York,1984; pp. 225-258.
[28] Wallsh F.C.; Gardner N.A.; Gabe D. R.; *J. Appl. Electrochem.*, 1982, 12, 299-309.

Index

T

V

W

X

Y

Z

About the Editor

Victor Martinez-Luaces, PhD

Victor Martinez-Luaces was born in Montevideo, Uruguay. He is a Chemical Engineer and Mathematician, and he also has a Master and a PhD in Education. He is a Researcher and Mathematics Consultant at the Chemical Engineering Institute, UdelaR (Uruguay), and a member of the PROFESTEAM Group, University of Granada (Spain). Additionally, he was a Visiting Professor at LDAR, University of Paris VII (France).

Dr. Martinez-Luaces was awarded twice with the first prize at RELME (2002, 2005) and received the Best Teacher Award at the Faculty of Engineering, UdelaR (1992, 1996). He was Chair, Convenor, and Keynote Speaker at International Scientific Conferences. He has published more than twenty books and book chapters (USA, UK, Germany, etc.), more than fifty journal papers, and served as a reviewer and guest editor in several international journals.